The Symbiosis of Work and Technology

Edited by

Jos Benders
University of Nijmegen

Job de Haan
Tilburg University

David Bennett
Aston University

UK	Taylor & Francis Ltd, 4 John St, London WC1N 2ET
USA	Taylor & Francis Inc., 1900 Frost Road, Suite 101, Bristol PA19007

British Library Cataloguing in Publication data

A catalogue record for this book is available from the British Library.

ISBN 0 7484 0316 7 (cased)
0 7484 0317 5 (paper)

Library of Congress Cataloging-in-Publication Data are available

Cover design by Amanda Barragny

Typeset by MHL Typesetting, Coventry

Printed in Great Britain by Burgess Science Press, Basingstoke, on paper which has a specified pH value on final paper manufacture of not less than 7.5 and is therefore 'acid free'.

Contents

List of Figures

List of Tables

List of Abbreviations

ABC	Activity Based Costing
ABM	Activity Based Management
AI	Artificial Intelligence
AMS	Advanced Manufacturing Systems
APS	Anthropocentric Production Systems
AuT	*Arbeit und Technik* (Labour and Technology)
BPR	Business Process Re-engineering
BU	Business Unit
CACSD	Computer-Aided Control System Design
CAD	Computer-Aided Design
CAM	Computer-Aided Manufacturing
CAPS	Computer-Aided Problem Solving
CE	Concurrent Engineering
CIM	Computer-Integrated Manufacturing
CNC	Computerized Numerical Control
DFA	Design for Assembly
DFM	Design for Manufacturing
DIN	*Deutsches Institut für Normung* (German Normalization Institute)
ETHICS	Effective Technical and Human Implementation of Computer-Based Systems
FAST	Forecasting and Assessment in Science and Technology
FMS	Flexible Manufacturing System
FT	*Fertigungstechnik* (Manufacturing Technique)
GALEN	Generalized Architectures Lexicons Encyclopaedias and Nomenclatures in medicine
GRAIL	GALEN Representation And Integration Language
HdA	*Humanisierung des Arbeitslebens* (Humanization of Working Life)
IAT	*Institut Arbeit und Technik* (Institute Labour and Technology)
ICAM	Integrated Computer-Aided Manufacturing
ICT	Information and Communication Technology
IDEF	ICAM Definition System
IMVP	International Motor Vehicle Program
ISO	International Organisation for Standardization
IT	Information Technology
JIT	Just-in-Time

KBS	Knowledge Based Systems
LP	Lean Production
MIT	Massachusetts Institute of Technology
MITOC	Management of Integrated Technical and Organizational Change
MST	Modern Sociotechnology
MTM	Methods-Time Measurement
NCMS	National Center for Manufacturing Sciences
NHS	National Health Service
NRW	North Rhine-Westphalia
NUMAS	Numerical Methods Advisory System
PEN&PAD	Practitioners Entering Notes & Practitioners Accessing Data
QWL	Quality of Working Life
REFA	*Reichsausschuß für Arbeitsstudien* (German Association for Ergonomics and Work Studies)
RHIA	*Verfahren zur Ermittlung von Regulationshindernissen in der Arbeitstätigkeit* (instrument to identify regulation barriers in industrial work)
SADT	System Analysis and Design Technique
SEM	Supportive Evaluation Methodology
SMART	Smart Manufacturing Techniques
SME	Small- and Medium-sized Enterprises
STEPS	Software Technology for Evolutionary Participative System Development
STS	Sociotechnical Systems
STSD	Sociotechnical Systems Design
TBCM	Team-Based Cellular Manufacturing
TPM	Total Productive Maintenance
TQM	Total Quality Management
VERA	*Verfahren zur Ermittlung von Regulationserfordernissen in der Arbeitstätigkeit* (instrument to identify regulation requirements in industrial work)
WOP	*Werkstattorientiertes Programmierverfahren* (Shop floor-oriented programming)
ZF	*Zahnradfabrik Friedrichshafen*

Foreword

As an edited collection, *The Symbiosis of Work and Technology* admirably pushes forward the frontier of research on 'human-centred systems'. In the last four decades a more complex understanding has emerged of the relationship between work and technology, whether between the social and the technical, or human factors and production systems. Such research flies under many different flags; 'anthropocentric', 'socio-technical systems', and so on. Their goals are broadly similar.

The *symbiosis* between these binary elements sets out to enhance organization design and system performance. A *symbiotic* approach has been presented as a *European* approach, for example (see Warner *et al.*, 1990, Chapter 1 references), although it has also found a niche in Scandinavia, the US and elsewhere, as the contributions to this book clearly set out. All are enthusiasts for the human-centred paradigm and their chapters link together very well.

Although earlier work at Volvo has now been substantially phased-out, renewed interest has recently emerged in 'high-tech' firms like Compaq computers. Several initiatives sponsored by the European Commission, such as in the ESPRIT and FAST programmes, have been a feature of the late 1980s and early 1990s. TQM has also played a part in pushing forward interest in 'empowerment' and similar forms of worker involvement. The novel term 'lean' socio-technology is introduced in this volume, for example. As competition in international markets increases, the search for new organizational designs to enhance technological potential has increased, in particular, to produce designs which combine quality and productivity, which is not a straightforward proposition at all. As Wobbe points out (p. 15), the more affluent Western consumers become, the more customized products are called for.

The international coverage of the volume must be noted, with contributions from both sides of the Atlantic, describing research from, for example, Australia, Western Europe, Japan, Scandinavia, the US and the United Kingdom, amongst other places. The search is for a 'humanistic, democratic and unstressful organization' (p. 35), as van Bijsterveld and Huijgen suggest. In the harsh business climate of the 1990s, these are very brave words indeed!

The influence of Japanese management has been a key feature of recent times. Interest in 'kaizen', as Karlsson notes (p. 56), has led to partial applications in Sweden. In Germany, too, the search for more flexible production systems has led to extensive organizational experiments, as Latniak outlines (p. 68), with programmes like *Arbeit und Technik*. Cost structures in his country are, however, becoming a preoccupation and humanistic work-experiments now fall on less receptive ears. In the Antipodes, sociotechnical change has been encouraged in team-based cellular manufacturing systems (TBCM), as Badham describes (p. 86), although how typical such innovations are there is hard to judge. In the US, new product and process technologies are *de rigueur*, with numerous CIM and JIT cases reported, such as by Majchrzak and Finley (p. 96). The recent sharp rise in American productivity levels may be associated with the diffusion of advanced technologies, although sociotechnical systems (STS) have not spread too rapidly, possibly because of their inherent complexity. One answer, then, is to reduce this dimension of the approach (p. 102). In artificial intelligence (AI) applications, we find a new challenge to human-centred systems. Kirby emphasizes 'empowered-user involvement in the design and development process' (p. 132).

Whether symbiotic approaches will become mainstream is hard to say yet. Benders *et al.* believe they are still in an 'incubation period' (p. 146). Buzz-words abound, 'empowerment' and 'de-layering', for example, but these do not effect organizational change in themselves. In fact, the Editors' conclusions are very cautious. The present writer can endorse their conclusions. It will only be by the demonstration of *results* that wider diffusion will be achieved. Businessmen and senior managers will only be impressed by 'bottom-line' justifications in these difficult times!

Professor Malcolm Warner
Fellow, Wolfson College, Cambridge
February 1995

Preface

The idea for a workshop and this book arose following the third biennial International Conference on Management of Technology (MOT) held in Miami in 1992, which was attended by the three editors. While discussing our previous work and current interests it emerged that, despite pursuing different areas of specialization, we could identify a common strand through the work that each of us was carrying out; namely the question of how to avoid treating the design of work and technology as two different issues. So was born the proposal to hold the workshop on 'Symbiotic Approaches to Work and Technology', which was organized at the fourth International Conference on MOT held in Miami early in 1994.

During the period between the two conferences a considerable amount of preparatory work was carried out. The workshop objectives were defined and prospective contributors contacted who were working in a variety of different areas relating to our general theme as well as representing a range of different countries. Before the workshop, we produced a 'position paper' which was sent out to the contributors and published in the conference proceedings. Each contributor then submitted a draft of their paper which was circulated to all the other contributors. The workshop occupied one full afternoon session of the conference during which time each contributor's paper was introduced and critiqued by one other specifically assigned contributor. A general discussion on each paper then followed. Every paper was discussed and every contributor prepared a critique of one of the papers.

Following the workshop, every contributor revised their paper in light of the comments that had been made. Each of these papers constitutes Chapters 2–8 of this book. At the same time, our own conference presentation was revised to take account of the specific ideas of each of the contributors together with the workshop discussions. Finally, the concluding chapter was written after we met in The Netherlands to review the contributions and agree our conclusions from the project.

As readers will appreciate, this book has been made possible by the efforts and co-operation of the individual chapter authors and workshop participants; we acknowledge and thank them. In addition, this project would not have been possible without the support provided by our institutions, the University of Nijmegen and Tilburg University in The Netherlands and the University of Aston in the UK. We must also acknowledge the role played by organizers of the Management of Technology conference. Finally, we are indebted to Richard Steele and his colleagues at Taylor & Francis.

Jos Benders
Job de Haan
David Bennett

September 1994

1

Symbiotic Approaches: Contents and Issues

Jos Benders, Job de Haan and David Bennett

Introduction

Many social scientists claim that the one-sided interest of technicians in technical aspects, which results in the neglect of social and organizational issues, leads to the creation and implementation of inadequately functioning systems. During the last 40 years, individual researchers and research groups have recognized the existence of this hiatus, and have tried to design production systems that were labelled 'sociotechnical' (Emery, 1959; Mumford, 1983), 'human centred' (Badham, 1991) or 'anthropocentric' (Rauner and Ruth, 1991; Wobbe, 1992). In the remainder the common denominator for such design methods is 'symbiotic', thereby stressing the necessity of cooperation between technical and social scientists in designing production systems. This new term indicates that neither technical nor social concerns dominate. The terms 'sociotechnical', human centred', and 'anthropocentric' all start with a word stressing the social, i.e. human element, which, although often against their original intentions, became dominant in many projects.

Although the importance of the symbiotic approaches has been stressed repeatedly during the last 40 years as a prescriptive device, it is striking that descriptive studies time and again demonstrate that these prescriptions seem to have contributed little to the design of new production systems in practice. Thus, the diagnosis that technical systems fail to achieve their goals due to a lack of consideration for organizational and social issues can hardly be called new. Repeatedly, researchers have pointed to insufficient performance of production systems because social, organizational and/or human factors were ignored when the system was designed. Despite this repeated diagnosis, the same problems occur over and over again (Perrow, 1983; Wall *et al.*, 1987; Majchrzak, 1988; de Sitter, 1989; Warner *et al.*, 1990; Alsène *et al.*, 1992; Freyssenet, 1992; Heming, 1992).

Perhaps this conclusion is too gloomy, but, even if true, that would be reason enough to pay attention to such methods. The approaches developed in the course of time need to be inventoried and studied, so that lessons can be learned in a systematic way from experiences of previous researchers. Furthermore, researchers currently working in this field are likely to gain from active interaction with colleagues. The workshop 'Symbiotic Approaches: Work and Technology' aimed to stimulate discussion between researchers in order to put the conceptual ideas and later the practical designs resulting from these conceptual ideas on a higher plane. Questions are piling up. If the above impression is correct, why have previous symbiotic methods not realized their promising potential? What actions need to be undertaken to create more favourable conditions for the implementation of contemporary symbiotic approaches? What are the basic differences and similarities between the different symbiotic methods? Is there anything 'new under the sun', or are these approaches basically the same?

Before even starting to discuss such questions, more insight is needed into the similarities and differences between the symbiotic approaches developed in different countries. Symbiotic approaches are applied at the level of technical systems, but also at higher levels of aggregation such as organization design. Whereas the first involves the incorporation of social criteria into the design of technical systems, the latter is concerned with characteristics that technical systems need to have in order to function effectively in the environment in which they are embedded. Furthermore, the process of elucidating the social criteria with which technical systems have to comply needs ample attention as well. These criteria often become known only when the design process is in full swing. The subsequent sections deal with these aspects in the following sequence: the design of (individual) technical systems, implications of organizational criteria, and the design process (each of these sections discusses two examples).

Symbiotic approaches

Technical systems

Arguably, the level of individual technical systems has been given most attention in the development of symbiotic methods. These often concentrate on the design of the man-machine interface, ergonomic requirements in general and the consequences for worker skills and learning. Many symbiotic systems have been developed in Germany (Scherff, 1989; Brödner, 1991; Hacker *et al.*, 1992), partly as a result of the stimulus given by substantial government funding in the form of the programme *Humanisierung des Arbeitslebens* (Humanization of Work) and its successor *Arbeit und Technik* (Work and Technology) (Herzog, 1990). Within the European Union many projects have been sponsored by the ESPRIT (Cooley, 1989) and FAST-programs (Gottschalch, 1990). Scandinavia also fulfils a prominent role with the UTOPIA-project (Bødker *et al.*, 1987) as an outstanding example. Whereas most researchers concentrate on manufacturing (Zarakovsky and Shlaen, 1991), information systems in general (Blackler, 1988) and knowledge-based systems (Kirby, 1992) have also been subject of research.

In 1989, the presentation of several programming packages under the name of *Werkstatt-orientiertes Programmierverfahren WOP* (shopfloor-oriented programming) drew much attention and aroused great enthusiasm. Traditionally, computerized numerical control-machines (CNC-machines) are programmed by writing abstract programs. This proved to be an insurmountable hindrance for many traditional craftsmen, who found themselves conducting standard tasks, such as loading and monitoring, after the introduction of CNC-machines. Such a change of job was often seen as inevitable because the intellectual requirements posed by the traditional way of programming was often perceived to be too high for traditional craftsmen. WOP meant to bring programming back to these craftsmen. With this purpose in mind, WOP has a user interface which presents objects and operations graphically. This allows (traditionally) skilled workers to use and maintain their existing knowledge. The sequence of operations is not determined automatically, but the worker remains in control. He is able to generate and evaluate alternative programs. Complex cutting forms are available in preprogrammed modules to assist the worker programming. The programs can be simulated and tested graphically. Tests by a leading German manufacturer indicated that

WOP was far more economical than existing programming packages, a finding which held for all degrees of parts complexity. Furthermore, the program can be mastered quickly (Brödner, 1991; Herzog, 1991).

A second example concerns a relatively new development — knowledge-based systems. As is generally the case when new technical systems are developed and gradually implemented, technical aspects predominate. There is evidence that organizational and social aspects, such as maintenance problems and degradation of work respectively, are neglected (Benders and Manders, 1993). Yet, human-centred knowledge-based systems are emerging, too (Danielsen, 1991; Kirby, 1992).

Kirby (1992) describes the development of a human-centred approach to the development of knowledge-based systems, which is called CAPS: 'computer assisted problem solving'. CAPS aims to overcome the disadvantages of traditional knowledge-based systems by considering three issues:

1. subject matter ('knowledge domain') of the advisory system;
2. the nature of the advice given by the system;
3. user control.

Unlike traditional systems, CAPS does not aim to replace a human (expert) by a technical system, but this system is seen as an instrument to support human decision-making: advice is given rather than binding commands. User control is guaranteed by using a hypertext system, which allows users to request data on a particular topic. Information on related topics, so-called 'buttons', is highlighted, and can be accessed as well. In order to avoid getting lost in an overload of information when selecting different buttons, it is necessary to formulate a problem-solving strategy into the organization of nodes and links in the system. So far, a prototype system has been developed to assist design engineers.

Organizational design

Technical systems as such are only one element of a production system. The performance of production systems partly depends on machinery, yet the people that work around and with these machines, and the structure of relationships between men and machines are equally important factors. Symbiotic design approaches claim to give these factors more consideration than conventional approaches.

The Swedes have long fulfilled the role of the world's leading symbiotic designers, with Volvo and Saab as key actors (Bennett and Karlsson, 1992). Volvo's Uddevalla plant has been a sociotechnical showcase, gaining worldwide attention. The design process took several years. The ultimate result was an assembly plant with six product shops, centred in groups of three around two inspection shops where the cars are tested. Every product shop contained eight parallel teams of approximately ten persons. These teams assemble complete cars in 2-hour cycles. Material supply for this new type of assembly plant demanded much attention, resulting in a highly automated system supplying the product shops from centrally located material shops. Ergonomic conditions received a great deal of attention too (Ellegård *et al.*, 1991).

The plant is claimed to have performed well, both in social and economic terms. The number of assembly hours per car was falling rapidly from an initially high level to a level comparable to Volvo's traditional Torslanda plant and below, a reputation for product quality was quickly built up, and its ability to handle a large product variety was impressive (Engström and Medbo, 1993). Despite this promising performance, Volvo decided to close the plant. A host of reasons has been reported, including a strong fall in the demand for cars as well as Volvo's future partner Renault's objection against such a sociotechnical plant.

In the field of organizational design, lean production has probably been the most hotly debated topic since 1990. Its proponents portray it as a hallmark of efficiency, whereas opponents point to its detrimental social effects (see the debate between Berggren (1994) on the one hand, and Adler and Cole (1994) on the other; Fucini and Fucini (1990)). However, around the beginning of 1990 the first signs appeared that lean production threatened to become the victim of its own success: the Japanese production capacity had expanded so grossly that the labour market had become extremely tight. Japanese car manufacturers faced serious labour market shortages, indicating that the Japanese workforce did not consider working in a lean production system as the most attractive employment opportunity. As sociotechnical solutions came under fire by lean production, scholars with a liking for sociotechnical systems design eagerly pointed to this new development. In particular, there was widespread interest in the claim that the conveyor belt had been questioned in the design of Honda's latest factory in Japan, which had been inspired by Volvo's sociotechnical factory in Uddevalla, Sweden (see Jürgens, 1993). The first indicative, yet empirical evidence (Sey, 1994) shows that Japanese car producers have adopted a whole range of measures to guarantee their staff levels. These include the introduction of ergonomic improvements, the creation of a pleasant physical work environment and, in some cases, the replacement of the conventional conveyor belt by automated-guided vehicles, which are placed, however, in a line layout. It is certainly too early to claim that lean production in its revised, 'post-lean' form can be labelled symbiotic.

The design process

The descriptions of the technical and organizational levels focus on structural characteristics. These can also be considered as the result of a design process, which is the subject of the following discussion. Whereas some symbiotic systems can be bought off the shelf, they will always have to be implemented in a specific work organization. More often than not, however, symbiotic approaches involve tailor-made solutions, requiring an intense and lengthy design process in which technicians and workers play key parts. Sustained managerial commitment to the project is vital. Whether or not such symbiotic projects need a structured design methodology is subject to debate, as eclectic approaches are used as well.

The ETHICS design methodology, meaning 'effective technical and human implementation of computer-based systems', was developed in the UK (Mumford, 1983). ETHICS has three goals:

1. To legitimize a value position in which future users at all levels participate in system design.
2. To enable design groups to take job satisfaction criteria into account, next to economic and technical criteria.
3. To ensure that a new technical system is embedded in a well functioning organization.

ETHICS has clearly been influenced by classical sociotechnical ideas as developed at the Tavistock Institute and by L. Davis. Participation of users in system design, job satisfaction and the integration of technology, people, tasks and the organizational environment form essential elements. A design group, preferably of 8–10 members, is to follow a systematic, 15-step procedure. It has been used by UK firms as well as a US manufacturer, leading to 'well designed "total" systems', as Mumford claims. Also, ETHICS does not require more man days in development than conventional methods, although its lead time is longer.

Depla (1988) describes a far more eclectic approach. Recognizing that organizations have a large freedom of choice with respect to factors affecting work design, he describes efforts by a militant Dutch union to change the traditional and pragmatic attitude toward work design as reflected in a plan with the prozaic name 'Slaughterline 2000'. This plan was designed to automate fully the process of slaughtering pigs, mainly by robotization. This plan's feasibility was questioned because of technical and economic reasons, leading to a more pragmatic alteration. Yet, what stayed was the neglect of attention for the quality of working life. This led the union to formulate an alternative, not as a 'definite counterplan made by social scientists that opposes the plans of the engineers' (Depla, 1988: 207), but to stimulate the discussion about possible alternatives, broadening the narrow technoeconomic scope. Yet, this alternative plan proved to be rather modest in its consequences for work design, focusing on job enlargement and the elimination of unfavourable working conditions. Nevertheless, Depla was pessimistic about the chances of realizing this symbiotic project. He judged that full automation might be more economic depending on the market strategy followed, and that it is even desirable in case of unattractive tasks. Furthermore, there is a trade-off between making work more attractive by paying attention to the quality of working life versus elimination of this work by automation. Finally, management's orientation to work design and automation and the union's power to influence this are essential factors for the chances to realize alternative plans.

Discussion

Despite the brevity of the accounts of various approaches given above, a variety of often interrelated issues for discussion can be distilled, whereby we play occasionally the role of the devil's advocate.

Degree of worker's participation

Some methods have a participative point of departure, i.e. opinions of workers are taken into account in system design. These methods stress the importance of job satisfaction, which is sought to be achieved by letting future users participate in system design. Other reasons for employee participation include system acceptance, input of employee knowledge and democratization. On the contrary, other methods have a more expert nature, i.e. experts in the field of symbiotic methods dominate system design. For instance, an objective standard is created by which to judge the quality of working life independent of the subjective account of employees.

Furthermore, how does democratization relate to symbiotic methods? Is user or employee input in the design process necessary because of democratization as a value *per se*, or are aspects such as making available necessary user knowledge to incorporate into the design and increasing the likelihood that users will accept the system when implemented, more important? What choices have to be made when the opinions of design experts, for instance engineers and consultants, are incompatible with the demands of future users, in other words: are expert and participant approaches reconcilable? To what extent does the possibility that conflicts occur necessitate a structured design process such as ETHICS?

Economic aspects

As Herzog (1991) states, unsatisfactory economic performance is a killer argument for the development and implementation of symbiotic approaches, in his case WOP. Not surprisingly, the more recent symbiotic methods, in particular, are claimed to have superior economic results, thanks to positive changes in workers' attitudes and behaviour, amongst other things. However, a major problem in this respect concerns quantifying these 'soft' benefits. Whereas the costs of the implementation of symbiotic projects (or any other project for that matter) are generally easy to establish, the benefits of their functioning can often only be established retrospectively. Thus, it may be hard to justify symbiotic methods by capital budgeting techniques, which require an *ex ante* assessment of a project's financial prospects, a problem which is aggravated in case of extensive and thus expensive design projects. Furthermore, the recent closure of Volvo's symbiotic showcase, namely its Swedish Uddevalla plant, has been criticized because of this plant's allegedly superior economic performance (Engström and Medbo, 1993). Although hard to accept for those who believe in rationalism, this would indicate that an adequate economic performance may not always be sufficient for symbiotic designs to last, and that deep lying resentments may play a role as well.

Finally, which organizations can afford the often long-lasting symbiotic design processes, which require substantial resources? Is this an impediment to the diffusion of symbiotic approaches in small- and medium-sized enterprises?

National environments

As Cole (1985) has shown for the case of small-group activities, the strands of employers' organizations, unions and (national) governments as well as labour-market conditions play a vital role in the diffusion of organizational concepts. Most symbiotic approaches stem from countries in Western Europe with Scandinavian countries and Germany fulfilling leading roles in the development. The southern countries of the European Community (Wobbe, 1992), and the United States of America and Japan have scarcely been involved (Rauner and Ruth, 1991), although the US sociotechnical tradition dates back from the 1950s. What are the major factors on the national level that respectively hinder or facilitate the development and diffusion of symbiotic methods? Consequently, is the competitiveness of symbiotic methods related to national environments? Is it a coincidence that Germany, with its emphasis on skilled labour and its strong system of vocational training, fulfils a leading role in the development of symbiotic systems? Should this be interpreted from an institutional point of view (see Maurice *et al.*, 1980), or is a culture oriented view (see Hofstede, 1980) to be preferred? Should they be seen as fitting in the specific West-European context characterized by small power distances and a tradition of participative and consensus-oriented relationships? Are symbiotic methods biased by Western values such as democratization/participation and autonomy?

The design process

Symbiotic methods stress the importance of simultaneous attention for technical, social and economic factors. System design has to take into account these factors simultaneously; neglecting this leads to sub-optimal designs. However, the common practice of muddling through, a term coined in 1959, does not seem to lead to large-scale bankruptcies, in other words: non-symbiotic approaches are often viable. Conventional sequential rather than symbiotic parallel decision-making with respect to technical, economic and social criteria has also led to acceptable results (de Haan *et al.*, 1992). Complex symbiotic design projects may deter management from using it, especially because management generally has a preference for simple solutions or even panacea (Gill and Whittle, 1993).

This, of course, may be because management is used to dealing with organizational changes in a specific way and will turn only under extreme circumstances to new or innovative organizational practices. This behaviour can be related to the notions of bounded rationality and organizational inertia.

'Wrong' use of symbiotic systems

Even industrialized countries are fairly dependent upon small- and medium-sized enterprises (SMEs). Typically, affairs are handled in an incremental way in SMEs and they are characterized by low levels of knowledge concerning organizational matters such as organization and job design. Symbiotic design projects are unlikely to be initiated by these firms; instead, they generally buy off-the-shelf systems. However, it is possible that a symbiotic system such as WOP may be used in a traditional way, facilitating programming not for a worker, but for a specialist programmer (Benders, 1993: 47).

Who stands accused?

It is sometimes claimed that conventional technical systems seduce organizations to implement fragmented jobs or even that they are developed consciously to create such jobs (Noble, 1978; Scarbrough and Corbett, 1992). At best, engineers are seen as being unaware of the social and organizational consequences of their technical designs. However, such claims, typically, are made by social scientists, accusing mechanical engineers. But these positions can be reversed: are social scientists sufficiently capable of specifying the technical requirements that symbiotic systems must meet? To what extent do social scientist need to have technical knowledge in order to communicate with technicians? Is it not too easy for social scientists to claim that engineers are insufficiently aware of social criteria while not possessing technical knowledge themselves?

If, during the last decades, descriptive studies repeatedly diagnosed that organizational and social aspects were neglected at considerable cost, why have prescriptive symbiotic approaches failed to be more effective? What factors impede the use of symbiotic methods, and is it possible to intervene?

Industrial setting

Frequently, symbiotic approaches are attributed best chances in case of unit and small batch production (Wobbe, 1992), indicating that symbiotic methods are primarily successful in particular market segments. In that case, mass production may retain a low quality of working life, whereas this sector needs improvements most from the perspective of the quality of working life. To what extent do chances for symbiotic approaches vary depending on market segments and output characteristics (Sorge, 1989)? What if economic assessments prove that conventional designs are superior? Are social and organizational needs always compatible?

Introduction of contributions

The main body of this book comprises the following seven chapters, each of which has been written by different authors who address the subject of symbiotic approaches from their own particular perspective.

The first contribution is from Werner Wobbe of the Commission of the European Communities. His paper is on 'Anthropocentric Production Systems' (APS) which have been conceived to cope with the problems of competitiveness and the quality of working life in traditional production systems. Against the background of changing market conditions and disappointing results of high-tech production systems he identifies the need for a new paradigm or *Leitbild* for the design of production systems, and discusses its prospects within the European Community, taking differences between member states into account.

Mark van Bijsterveld and Fred Huijgen of the University of Nijmegen in The Netherlands wrote the second contribution. They introduce the idea of 'Modern Sociotechnology' (MST), a special variant of symbiotic design. The main characteristics of MST are described and its applicability and (economic and social) performance in different cultural and institutional environments and production situations is considered. After discussing lean production and its impact, they turn to the weaknesses of modern sociotechnology and, finally, they end their contribution with a number of conclusions, among which is the feasibility of a 'lean sociotechnology'.

Ulf Karlsson of Chalmers University of Technology provides the next contribution. He plots the history of the Swedish sociotechnical approach and considers some of the problems of conducting design research. Karlsson's contribution reviews some recent developments relating to technology analysis for R&D organizations, engineering work design and accounting practice. Swedish and Japanese management are then compared and consideration given to how combining both can lead to broadened concepts of job design.

Erich Latniak, of the Institut Arbeit und Technik, gives an overview of German experiences on *Technikgestaltung* (shaping of technology). Although the importance of the organization and the workforce's competence for competitiveness are now widely acknowledged, symbiotic approaches are still not widely diffused. The main problem is the lack of an integrated approach for the restructuring process. He stresses that there is no coherent approach shaping technology and organization but there are manifold methods and tools that must be adapted to the specific needs and problems of factories.

Managing sociotechnical change is the subject of the chapter by Richard Badham of Wollongong University in Australia. Badham introduces the 'configuration process approach' which stresses that specific techniques are malleable and are adapted to suit the needs of a particular organization and its members. This intensely political process is illustrated by a case study on the 'smart manufacturing techniques' (SMART) project.

The next chapter is by Ann Majchrzak and Linda Finley of the University of Southern California and Texas Instruments, respectively. They discuss the sociotechnical systems design as practised in the USA, stressing that its abstractness and complexity are two major dilemmas next to the fact that trade-offs need to be made continuously. The solution for these dilemmas is found in a software tool called 'ACTION'. A case is described to illustrate the practical use of their approach.

The final contributor is John Kirby of the University of Manchester. Like the previous chapter this is more focused. It first describes an application of the basic human-centred philosophy to the design of computer-aided design for control engineers. It then goes on to present and discuss a project, in which a user-centred approach to design has been developed. Next, some implications are considered for the human-centred approach in terms of the purpose of systems, the design process and knowledge representation.

Despite the obvious differences in the detailed content of these seven contributions they do, nonetheless, each constitute a perspective on the broad concept of symbiotic approaches. Based on these perspectives, the final chapter concentrates on the question of whether such approaches will become 'mainstream'.

References

Adler, P.S. and Cole, R.E. (1994) Rejoinder, *Sloan Management Review*, **35** (2), 45–50.

Alsène, E., Lefebvre, L. and Auclair, S. (1992) The global approach in the management of technological change: An assessment, in: Khalil, T.M. and Bayraktar, B.A. (Eds), *Management of Technology III; The Key to Global Competitiveness*, Norcross: Industrial Engineering and Management Press, pp. 1364–1374.

Badham, R. (1991) Human-centred CIM; Informating the design-manufacturing interface, *Futures*, **13** (10), 1047–1060.

Benders, J. (1993) *Optional Options: Work Design and Manufacturing Automation*, Aldershot: Avebury.

Benders, J. and Manders, F. (1993) Expert systems and organizational decision-making, *Information and Management*, **25** (4), 207–213.

Bennett, D. and Karlsson, U. (1992) Work Organization as a Basis for Competition; The Transition of Car Assembly in Sweden, *International Studies in Management and Organization*, **22** (4), 49–60.

Berggren, C. (1994) NUMMI vs. Uddevalla, *Sloan Management Review*, **35** (2), 37–45.

Blackler, F. (1988) Information technologies and organizations: Lessons from the 1980s and issues for the 1990s, *Journal of Occupational Psychology*, **61** (2), 113–127.

Bødker, S., Ehn, P., Kyng, M., Kammersgaard, J. and Sundblad, Y. (1987) A Utopian experience; On design of powerful computer-based tools for skilled graphic workers, in: Bjerknes, G., Ehn, P. and Kyng, M. (Eds), *Computers and Democracy*, Aldershot: Avebury, pp. 251–278.

Brödner, P. (1991) Design of work and technology in manufacturing, *International Journal of Human Factors in Manufacturing*, **1** (1), 1–16.

Cole, R.E. (1985) The macropolitics of organizational change; A comparative analysis of the spread of small-group activities, *Administrative Science Quarterly*, **30** (4), 560–585.

Cooley, M. (1989) Human-centred systems, in: Rosenbrock, H.H. (Ed.), *Designing Human-centred Technology: A Cross-disciplinary Project in Computer-aided Manufacturing*, London: Springer-Verlag, pp. 133–144.

Danielsen, O. (1991) *Human Centredness and Expert Systems*, FAST occasional paper 268, Brussels: Commission of the European Communities.

Depla, M. (1988) Automation in the Dutch meat industry: Slaughterline 2000?, in: Buitelaar, W. (Ed.), *Technology and Work; Labour Studies in England, Germany and the Netherlands*, Aldershot: Avebury, pp. 195–217.

Ellegård, K., Engström, T. and Nilsson, L. (1991) *Reforming Industrial Work — Principles and Realities*, Stockholm: Arbetsmiljöfonden.

Emery, F.E. (1959) *Some Characteristics of Socio-technical Systems*, London: Tavistock Institute.

Engström, T. and Medbo, L. (1993) Intra-Group Work Patterns in Final Assembly of Motor Vehicles, in: Karlsson, C. and Voss, C. (Eds), *Management and New Production Systems; The 4th International Production Management Conference*, London: London Business School/EIASM, pp. 395–410.

Freyssenet, M. (1992) Processus et formes sociales d'automatisation; Le paradigme sociologique, *Sociologie du Travail*, **34** (4), 469–495.

Fucini, J.J. and Fucini, S. (1990) *Working for the Japanese; Inside Mazda's American Plant*, New York: The Free Press.

Gill, J. and Whittle, S. (1993) Management by panacea: accounting for transience, *Journal of Management Studies*, **30** (2), 281–295.

Gottschalch, H. (1990) Produktionsarbeit in CIM-Strukturen, *Technische Rundschau*, **82** (26), 28–39.

de Haan, J., Peters, R. and Giesberts, A. (1992) The implementation of advanced manufacturing techniques: experiences of six Dutch factories, in: Hollier, R.H., Boaden, R.J. and New, S. (Eds), *International Operations; Crossing Borders in Manufacturing and Service*, Amsterdam: North-Holland, pp. 215–220.

Hacker, W., Hallensleben, K. and Teske-el Kodwa, S. (1992) Menschzentrierte CAM-Technologien ohne Reorganisation? Eine Studie zur Arbeitssystemgestaltung, *Zeitschrift für Arbeitswissenschaft*, **46** (3), 155–159.

Heming, B.H.J. (1992) 'Kwaliteit van arbeid, geautomatiseerd . . .; Een studie naar kwaliteit van arbeid en de relatie tussen automatisering, arbeid en organisatie', unpublished PhD dissertation, Delft University of Technology.

Herzog, H.-H. (1990) *Arbeit und Technik; Chancen und Risiken für die Arbeitswelt von Morgen*, Bonn: Projektträgerschaft Arbeit und Technik.

Herzog, H.-H. (1991) WOP auf dem steinigen Weg zur Norm, *Technische Rundschau*, **83** (22), 48–53.

Hofstede, G. (1980) *Culture's consequences. International differences in work-related values*, London: Sage.

Jürgens, U. (1993) Mythos und Realität von Lean Production in Japan, *Fortschrittliche Betriebsführung und Industrial Engineering*, **42** (1), 18–23.

Kirby, J. (1992) On the interdisplinary design of human-centered knowledge-based systems, *International Journal of Human Factors in Manufacturing*, **2** (3), 277–287.

Majchrzak, A. (1988) *The Human Side of Factory Automation*, San Francisco: Jossey Bass.

Maurice, M., Sorge, A. and Warner, M. (1980) Societal differences in organizing manufacturing units: A comparison of France, West Germany, and Great Britain, *Organization Studies*, **1** (1), 59–86.

Mumford, E. (1983) *Designing Human Systems for New Technology: The ETHICS Methods*, Manchester: Manchester Business School.

Noble, D.F. (1978) Social choice in machine design; The case of automatically controlled machine tools, and a challenge for labor, *Politics and Society*, **8** (3/4), 313–347.

Perrow, C. (1983) The organizational context of Human Factors engineering, *Administrative Science Quarterly*, **28** (4), 521–541.

Rauner, F. and Ruth, K. (1991) *The Prospects of Anthropocentric Systems: A World Comparison of Production Models*, FAST occasional paper 249, Brussels: Commission of the European Communities.

Scarbrough, H. and Corbett, J.M. (1992) *Technology and Organizations; Power, Meaning and Circuit*, London: Routledge and Kegan Paul.

Scherff, B. (1989) Ergonomischer Einsatz von Sprache als Teil der Mensch-Maschine-Schnittstelle zur Programmierung von CNC-Robotern für das Schweißen, *Zeitschrift für Arbeitswissenschaft*, **43** (4), 224–233.

Sey, A.-P. (1994) 'Soziale Aspekte in den gegenwärtigen Modifizierungen von

Produktionskonzepten in der japanischen Automobilindustrie; Japan auf den Weg zur "arbeiterfreundlichen" Fabrik?', unpublished Master's thesis, Free University Berlin.
de Sitter, L.U. (1989) *Modern Sociotechnology*, Den Bosch: KOERS.
Sorge, A. (1989) An essay on technical change: Its dimensions and social and strategic context, *Organization Studies*, **10** (1), 23–44.
Wall, T.D., Clegg, C.W. and Kemp, N.J. (Eds) (1987) *The Human Side of Advanced Manufacturing Technology*, Chichester: Wiley.
Warner, M., Wobbe, W. and Brödner, P. (Eds) (1990) *New Technology and Manufacturing Management; Strategic Choices for Flexible Production Systems*, Chichester: Wiley.
Wobbe, W. (1992) 'Anthropocentric Production Systems: Shall the EC intervene in the new emerging structures of European Business?', paper presented at 'Autonomy and Independent Work' workshop, Nijmegen, 1 December.
Zarakovsky, G.M. and Shlaen, P.Y. (1991) Ergodesign of flexible automated manufacturing psychological aspects, in: Roe, R.A., Antalovits, M. and Dienes, E. (Eds), *Technological Change Process and Its Impact on Work*, Budapest: Hungarian Coordination Council for Work Psychology/European Network of Organizational and Work Psychology, pp. 105–112.

2

Anthropocentric Production Systems: A New *Leitbild* for an Industrial Symbiotic Work and Technology Culture in Europe

Werner Wobbe

The call for a new *Leitbild*

A paradigm shift in the use of technology and labour and its organization has emerged recently, putting more emphasis on organizational factors than on technology in order to raise productivity. Until then, manufacturing concepts have been derived mainly from mass production, based on automated facilities, ameliorated by information technology. These concepts include hidden assumptions and *Leitbilder* (German: *Leiten* = 'to lead', *Bild* = 'image', 'picture') about how to produce. The underlying assumptions control the thinking and beliefs of designers and industrial managers about technology and work organization. In this sense they form a kind of industrial engineering culture. At its core, the underlying assumptions and Leitbilder derived from an American context are detrimental to the basic industrial structure of Europe, which is composed mainly of small and medium-sized enterprises, and hinders the adaptation process to a modern economy.

During the last decade, a market shift in customized quality products has grown at the cost of standardized mass products. Industries based on mass products and price competition have lost out to the Far East and might in future face competition from Eastern Europe. In Western Europe, the increase in product variants, quality features, small batch sizes and the decrease of the product life-cycles and repeat orders has dramatic consequences for management and organization in the manufacturing industries. During the 1980s, it has become evident that the demand flexibility could not be met only by new technologies.

In addition, the problem of technological complexity has called for *symbiotic* approaches to handle production appropriately. The old view of automation is blind to the fact that the more advanced and more complex manufacturing becomes, the more valuable will be the human skills and the human aspects of organization. Therefore, the future of advanced manufacturing might be centred around new factory and work organization patterns and around the human resource question and adapted technologies. In consequence, new design rules and new manufacturing Leitbilder have to be developed in order to cope with developments in advanced manufacturing. Leitbilder represent a set of core beliefs and assumptions which have become common knowledge to social groups or even the general public. A new Leitbild, adapted to the realities of the new European economy, is in the course of being created.

The so-called 'anthropocentric production systems' (APS) are at the core of this reasoning. They have been conceived to cope with the new world-market challenges and to overcome the problems reported at the end of the 1980s by advanced industrial countries about inefficiencies, failures and accidents in high-tech production systems. APS are built on the knowledge and skill of the work force and the collaboration and responsibility of the different groups, teams and departments which are aided by adapted technologies. The APS concept stretches from plant organization, via departmental cooperation, group work to the work place. To achieve a new production system, it might start as a single attempt which can be extended to the complete system.

In view of the Japanese challenge, the MONITOR-FAST programme of the European Communities called for developing anthropocentric production systems in order to modernize fundamentally European industries. However, the main obstacle to be overcome in this process will be the attitudes of the industrial actors educated of the Leitbild of mass production.

The issues mentioned above are dealt with more elaborately in the remainder of this contribution.

Reconsidering the basics in manufacturing

A change in the engineering orientation from the dominance of technical questions in manufacturing to reconsider the work organization has come late in the 1980s. The 'soft factors' in manufacturing have grown at the expense of the hard technological ones. In the scientific debate *Das Ende der Arbeitsteilung* ('The end of the division of labour?') by the German scholars Kern and Schumann (1984) might have been a starting point of detecting problems in the computer-integrated manufacturing (CIM) orientation.

Developments in the 1970s and 1980s

Expanding markets and world competition have led to enforced division of labour in this century and to taylorist and fordist principles of work organization. Achieving maximal scale of production via specialization and automation were the guiding concepts for managers. Increased labour specialization and automation have been complementary processes.

When, during the 1970s and 1980s, fragmentation of markets and product complexity grew and production flexibility became a threat, the computer-aided technologies seemed to offer a solution to cope with this challenge in manufacturing. Therefore, in these two decades the technological imperative, the technocentric manufacturing Leitbild, became dominant in managers' minds and in public support for R&D. At the end of the 1980s, the first critical conclusions were made by researchers observing the diffusion pattern of the computer-aided technologies;

- industrial robots, the symbols of industrial automation, are far beyond expectation in industrial development (Deiß *et al.*, 1990);
- flexible manufacturing systems (FMS) are hardly economic if applied in tayloristic forms of work organization (Haywood and Bessant, 1990);
- information technology in manufacturing had been applied very differently according to social constellations (Campbell *et al.*, 1989);

- in the Danish context, a paradox in increasing technology application and diminishing productivity has been observed (Gjerding *et al.*, 1990);
- in Japan the capital productivity has always been an issue complementary to organizational aspects (Kageyama, 1993).

Therefore it can be stated that at the beginning of the 1990s the manufacturing concepts inspired by information technology aiming at the unmanned automated factory have been put into question.

At this time, the complex interplay between man, technology and organization and its management has been recognized as a system. As a transitory phase, measures that foster the acceptance of new technologies have been of crucial interest. The training needed to adapt people's competence to new technologies was seen as instrumental. However, the old-fashioned tayloristic concepts of manufacturing organization has been kept.

Now in the 1990s, influenced by the MIT study about 'lean production' (Womack *et al.*, 1990) and the EC on APS (Lehner, 1991; Wobbe, 1991), it has become clear that the manufacturing organization, the use of skilled and competent labour and a proper management of this system are at the core of the solution for problems witnessed. Therefore, the manufacturing Leitbild is in the course of shifting from its technocentric focus to one that is anthropocentric and collaborative.

The background for the shift

As already mentioned, the understanding of manufacturing through the glasses of automation technology as well as the restricted view of informatics, searching for problem solutions in mathematical wisdom, is evidently insufficient. These 'biased views' of manufacturing are stemming from academic groups and also from military based research and manufacturing developed in the United States of America. For European manufacturing, it is relevant to understand that product markets are under transformation. This market shift is the economic background for the success or failure of manufacturing concepts.

The more affluent the Western societies have become, the more customized products are demanded. Quality features have come more and more into play, not only that of reliability but also those of additional functions and of design. While the cheap and price competitive products are increasingly produced in low wage countries, the markets for customized quality products are expanding throughout Europe. This increase of the customized quality economy has a high impact for manufacturing. The example of a large British electro-mechanical producer may serve as an illustration of developments in industry, as shown in Figure 2.1.

All efforts of the corporation to reduce the variants had not been successful. On the contrary, the product variants have steadily been increased, and at the same time the demand for quality upgrading has increased in parallel. On the other hand, the life-cycles of products have decreased and with them, the repeat orders. Also, the batch sizes have been reduced dramatically due to the demand of customers, to minimize stocks, and deliver the right portions when needed. In other manufacturing industries, and even in chemicals, this trend is observed.

These contradicting trends in manufacturing, a dilemma in the philosophy of price competition, can only be handled by a different way of manufacturing than that of applying mass production principles.

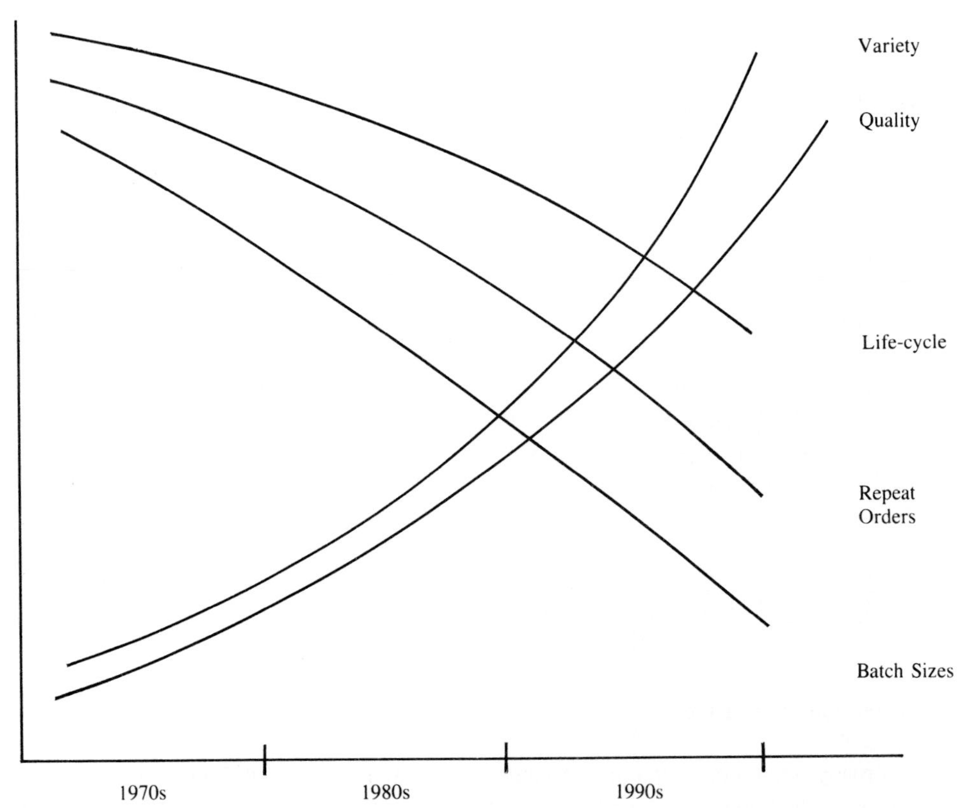

Figure 2.1 Market trends 1970–1990s. Source: Adapted from Kidd, 1990, p.5. Reprinted by permission of the FAST-programme of the Commission of the European Communities

APS: The new *Leitbild* for management, labour and organization

The anthropocentric production systems (APS) have been developed in order to cope with the customized quality economy. In general terms, their principles or their Leitbild in organizing the manufacturing process could be described in a nutshell as follows:

- comprehensive use of human abilities and performance;
- permanent learning of the work force and facilitating corporate structures;
- decentralized production units;
- cut-backs in the division of labour;
- collaborative forms of organization;
- adapted technologies.

In order to make these principles more concrete, their implications for designing production systems are made clear in the following subsections, which deal with various levels within an organization and the technological requirements.

The organizational design of APS on different levels inside the firm

At the factory level the guiding principle of APS points to decentralized production units. Plants can be organized into different product shops and be regarded as companies within a company. The main guiding principle would be the delegation of responsibility to lower levels to provide more autonomy.

The interdepartmental relationships and their collaborative functions are of central importance. In the course of growth of industrial enterprises, a strong departmentalization has occurred. Management has been confronted with coordination problems because departments have acted as separate entities. This separation has consequently created limitations to productivity, integration of firms, and led to an unproportional growth of indirect labour.

The Leitbild for APS is to increase day-to-day cooperation between 'experts' at all levels: the experts in executing the shop-floor work and the experts who have planned this work; the experts who make the decisions and give the orders and the experts who have asked for the order; the experts who run the machine and the experts who programme it. In practice, this means the instigation of face-to-face dialogue between designers, planners and manufacturing workers. This would lead, for example, to an interactive programming process between technician and worker on very complex parts, or the intensification of collaboration and early agreements between the business departments and the shop-floor concerning planning and scheduling. Cooperative relationships between white-collar and blue-collar workers are central. Barriers between these groups which are particularly entrenched in the Anglo-saxon industrial context and in some Mediterranean countries need to be abolished. These barriers do not exist in Japan and are much less prevalent in Scandinavian countries (see Karlsson, Chapter 4) and Germany (see Latniak, Chapter 5).

The working-group level might not even exist in strong hierarchical organizations. Therefore, one of the central innovations of the last two decades in manufacturing has been the introduction of working groups in the automobile industry as well as in mechanical engineering or chemical plant supervision and maintenance. The case of Volvo has become the most famous example with its semi-autonomous assembly groups. These organizational developments have given rise to completely new concepts of logistics (Auer and Riegler, 1990). Other cases are working groups with a low division of labour in highly automated systems, for example in FMSs (Hirsch-Kreinsen *et al.*, 1990).

Another case is the so-called 'production island' where discrete parts of manufacturing are grouped around a family of parts. The underlying principle in all these cases is the transfer of a broad range of functions, occupations and decision space to the group and to ensure that most members can cope with it (Brödner, 1985).

The APS design for the work place demands a broad range of measures to ensure skilled and collaborative work. The work place design has not only to create optimal work conditions but also to provide an environment which stimulates innovation, learning and collaboration based on a range of competence. Besides the design of the work organization, the man/machine relationship is challenged by software design.

In more detail, in highly automated areas the integration of job tasks, i.e. programming, scheduling, maintenance, processing etc. are central. In the assembly areas the work enrichment towards hybrid assembly is a way of work design. In lowly automated areas of batch production, work enrichment by planning, scheduling and maintenance tasks, as well as job rotation, are possible actions which can be taken.

APS requirements for technology

In the same way as existing examples of 'living' implemented APS organizations can be found (Brandt, 1991), there already exists pieces of APS technology. For example, the

Table 2.1 *Leitbild* for APS organization at different plant levels

Levels	Principles
Factory	* small decentralized production units * product shop * companies within a company * delegation of responsibility to lower levels
Interdepartmental relations	* cooperation between design and manufacturing * interactivity between workshop and engineering departments concerning programming of machine tools * integration of business department, technical department and shop-floor concerning planning and scheduling
Group	* installation of production islands * group work in Flexible Manufacturing Systems (FMS) * semi-autonomous assembly groups
Work place	* workshop programming * integration of intellectual and manual functions * in highly automated areas: integration of programming, planning, maintenance and processing tasks * in less automated assembly areas: work enrichment with decision on space on execution sequence and performance.

Source: Wobbe, 1991, p.22. Reprinted by permission of the FAST-programme of the Commission of the European Communities

shop-floor programming package 'WOP', developed in the German programme 'Fertigungstechnik' is already on the market (also, see Latniak, Chapter 5) The same holds true for the design sketchpad and the group-supporting scheduling system. Although there are concrete examples of software and hardware for APS technologies, there is still a broad scope for technological development. The core of these APS technologies is to be found mainly in the software component and its potential to aid APS structures. In contrast, the hardware is the less important part in APS technologies.

It has to be borne in mind that APS technology alone does not guarantee APS structures. These have to be achieved by organizational measures, and steps towards APS can even be achieved without APS technology. Thus, the technology has to be considered merely as a tool to work with. It is important to allow for collaborative workshop programming and interactive communicative tools with the design and technical office, but the technology alone is no guarantee that such working patterns will actually occur. Finally, it has to be assured that the technology does not block organizational measures, but aids them. New technology has to be developed for the full potential of APS to be realized.

Table 2.2 lists APS technology and technologies which can be used in APS structures, as well as the research and development areas which should be realized. Research and development are needed to elaborate 'tools' applied in the workplace which are IT-based but equipped with an analogue user surface, which is highly transparent for information, decision and control purposes within the manufacturing process. The man/machine interface has to work with elaborated symbolic representations which can be composed to make up complete pictures. Therefore, new vision systems and adaptable natural language interfaces are required to support the analogue and to guide the working behaviour of the user. The devices should also include learning supports concerned with the systems as well as with the working process.

Table 2.2 APS technology

Level	Technology
Work place	* shop-floor programming systems for machine tools and robots * decision support systems * analogue user support mechanism to control the manufacturing process * symbolic representations of complete pictures for information, processing and decisions * skill supporting and learning techniques
Group	* scheduling and planning systems for group work * computer aided cooperative work techniques for information, planning and decisions
Interdepartmental	* IT systems to facilitate interactions and dialogue between office and shop-floor * transportable analogue design sketch pads
Factory	* information systems to support network organizational structures
General	* adaptable and natural language human/computer interfaces * highly transparent support systems for collective and individual decision making * new vision and symbolic representation systems

Source: Wobbe, 1991, p.24. Reprinted by permission of the FAST-programme of the Commission of the European Communities

At the collaboration level between different departments or groups, new software systems have to be developed. They will aid the collaboration through information and decision support, rendering the overall cooperation process transparent and thereby permitting simulation and interaction.

Industrial culture in Europe and APS

Considering APS as an advanced form of manufacturing adapted to the upcoming customized quality economy, the prospects for a broad diffusion in Europe are unfortunately not too bright. Obstacles and hindrances can be observed in the traditional Leitbild of mass production and in traditional hierarchical structures in the manufacturing. In contrast, for APS collaborative structures are essential.

In the last few years, the knowledge about fundamental constellations of industrial cultures has been increased (Rauner and Ruth, 1991). Variations of industrial cultures have been distinguished between Japan, Europe and the USA, and even within Europe these constellations are very different (Lane, 1989).

Dimensions of collaborative structures

The traditional manufacturing paradigm has worked well with a hierarchic organization, high division of labour and clear task description. The new manufacturing organization needs overall collaboration in order to adapt steadily to new requirements and changes for the product. Collaboration is demanded at different levels and situations and requires:

- team work at all levels of production;
- blue/white-collar communication;
- interdepartmental cooperation;
- collaborative industrial relations;
- cooperation in the supply chain.

These requirements touch attitudes, values, the behaviour, social distinction, power and influence in corporations. They are the core of a given industrial culture — how people and organizations deal with each other — and touch immediately the management of a firm.

Taking into account the difficult social change process, the question is where does Europe stand with the introduction of organizationally advanced manufacturing. Many Europeans firms are modernizing their production on the lines of APS, particularly leading firms (Brandt, 1991). However, the overwhelming majority has not been affected by modern developments. Also, a split can be observed between advanced and less advanced manufacturing countries in Europe, pointing to the danger of widening the welfare gap between those countries.

The introduction of advanced manufacturing in the European Union

The MONITOR-FAST programme has carried out research on the assessment of the socioeconomic prospects of organizationally advanced manufacturing systems (AMS) in the European Union. The results are as follows (Lehner, 1991):

- The development towards advanced systems in the Member States is slow and uneven.
- The introduction of new computer-aided production systems follows predominantly traditional production concepts putting the focus on the technology and keeping a high division of labour.
- Anthropocentric production systems are, however, developed in an increasing number of advanced implementations in different industries and different types of firms.
- Considerable differences are observed in industrial sectors in the development of APS. Generally speaking, development of APS is stronger in more competitive and technologically sophisticated industries.
- Differences between the core industrialized countries are, for the most part, gradual rather than fundamental. Compared with other industrialized countries in the European Union, Germany is more advanced with respect to work organization, but not with respect to firm organization. Developments in the United Kingdom and Belgium are weaker than in other industrialized countries.
- In the less industrialized countries of the European Union (Greece, Ireland, Portugal and Spain), the advanced manufacturing systems are generally weaker than in most of the industrialized countries. In Ireland and Spain, there are, however, already experimental approaches to APSs, and a few examples exist in Portugal.

Obstacles for APS in industrialized countries

Obstacles to the modernization of industry towards APS are observed not only in the less industrialized member countries but also in the technical infrastructures of the industrialized countries:

- Modern computer-based production technology can be associated with a range of alternative solutions concerning the design of jobs, work and firm organization, qualification and incentive structures. More important impediments and obstacles to the development of APS are associated with social and economic structures.
- In all countries of the European Union, management strategies concerning the application of computer-based production technology are still predominantly orientated towards tayloristic production concepts.
- In some countries, manufacturing is still strongly characterized by standardized mass production while flexible, customer-oriented production is weak.
- In all countries of the European Union, rigidities in organization, status systems and wage structures are widespread factors hindering the development of APSs.
- In many countries, e.g. in the United Kingdom, Spain and Portugal, shortages of skilled labour exist due to weak vocational training and further training.
- In some countries, especially in the United Kingdom, France and Italy, low-trust industrial relations impede the redesign of jobs and organization.

The most important factors which might prevent the modernization of manufacturing industry lie in the lack of awareness of the benefits, and a lack of knowledge and experience which have, up to now, not led to an 'APS culture' which can continue and further develop on its own. These qualitative indications can be made more concrete for three countries: France, Germany and the United Kingdom.

National manufacturing cultures in three countries

With the help of a basic sketch of an enterprise, the composition of an organization's workforce can be analysed. Besides the manager, the white-collar staff consists of technicians, engineers, supervisors and clerical, commercial and administrative employees. The blue-collar workers conduct production work, maintenance and tool making.

As Figure 2.2 shows there are quite different proportions in Germany, France and the United Kingdom in the composition of white/blue-collar staffing (Lane, 1989). While France employs on average 41.6 per cent white-collar employees and 58.4 per cent blue-collar workers in manufacturing firms, Germany is significantly different. It has a white/blue-collar ratio of 28.2 per cent to 71.8 per cent, while the UK is closer to France with 37 per cent to 63 per cent. To execute work at the lowest possible competence level, therefore, it is best done in Germany, which might be explained by its traditional highly skilled workers.

The collaborative structures and the distinction between departments and areas are also considerably different between the countries. Because of the trade union structure based on professions, the distinction in the UK between the technical staff and the supervisory staff, as well as between production and maintenance workers, is strong. It is again different in Germany where this distinction is blurred because of the industrial trade union organization as well as the acceptance and the esteem of practical and productive functions in the German society.

In conclusion, out of these three countries, the German industrial cultural context might be best suited for the application of new collaborative production concepts, and it might explain its industrial strength in certain industrial sectors, such as machine tools, automobiles or electromechanical engineering.

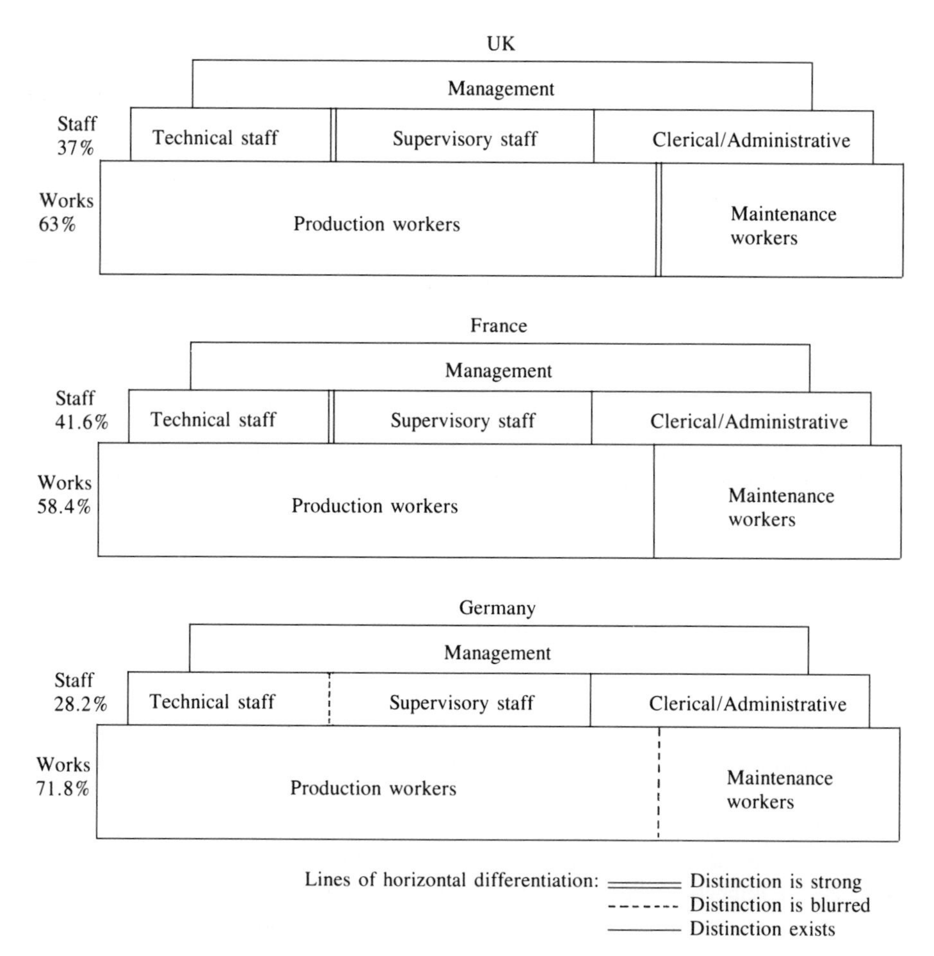

Figure 2.2 National organizational configurations. Source: Lane, 1989, p.47. Reprinted by permission of Edward Elgar Publishing Ltd

The prospects for the symbiotic approach of APS

Prospects in terms of rapid industrial modernization in Europe towards APS might be not bright in the immediate future due to the national manufacturing cultures which have a strong retending element. Nonetheless, there are already many advanced examples on hand.

The role of the European Union

In view of world-market pressures on the European economy, action is demanded from the European Public Authorities. This has to be stated, although industry has to solve manufacturing problems itself and in the light of 'socio-cultural' aspects of manufacturing and the spirit of subsidiarity in a double sense: the state should not touch the industrial firms' autonomy and the European Union should not hurt the autonomy of the national industrial relations.

Nonetheless, Member States should be made aware of the large differences between nations in the production modes in order to maintain a decent competitive international level. This would mean that the European Union has to play a role in communication, knowledge transfer and stimulation, but not in direct intervention.

According to the study on APS, action demanded from public authorities consists of (Wobbe, 1991):

- improving awareness of new manufacturing concepts;
- education, training and curricular reforms of universities;
- research in manufacturing organization and adapted technologies;
- knowledge transfer for less industrialized member countries;
- stimulation of collaborative cultures;
- fostering rapid diffusion of APS practices by demonstration projects.

These actions would enrich the new industrial policy of the European Union to strengthen the infrastructure, not only in its 'hard', but also in its 'soft side' which is the industrial culture. A first and fundamental step might be to work for a Leitbild change in Europe in the engineering community in order to build the proper foundation for further development.

The call for a Leitbild change towards an anthropocentric production culture

The traditional engineering Leitbild for organizing manufacturing has been quite successful and has not been challenged for a long time. Therefore, it has become technology centred on technocentric. In the old fashioned spirit the aim of designing manufacturing systems was to make the IT-based technical systems active to the exclusion of the human user, thereby rendering him passive (Cooley *et al.*, 1990). The complex systems should work in a predictable and respectable manner. All the flexible, innovative and adaptive abilities of human employees, such as subjective judgement, tacit knowledge intentionality and intuition are seen as disturbing factors which do not fit into the master plan and the Leitbild of designers, engineers and managers. In certain professional communities, these Leitbilder have become core beliefs and assumptions of common knowledge. As such they are taken for granted and are not seen as problematic.

These technocentric Leitbilder are fostered by white-collar employees, who maintain the belief that manufacturing problems can only be solved by mathematical and engineers' knowledge. They assume manufacturing challenges and increased productivity will be met by technology to which operators have been adapted. White-collar workers believe in the superiority of university education, that does not necessarily need manufacturing experience, but does need mathematical logic. In this way, an entire country's culture can be influenced by these assumptions. The belief is that the blue-collar occupations need less education, just a certain degree of training for the job. To safeguard white-collar authority, a hierarchical enterprise structure and centralized control is necessary.

The anthropocentric approach in society, by contrast, is guided by an acceptance and fostering of blue-collar work expertise. This approach depends broadly on educated human capital at all levels of the manufacturing chain. It aims at productivity achievements by organizational means, by personal commitments and decentralized patterns of control and cooperation. Only this kind of manufacturing organization can sustain world competition with flexible quality production, steady innovation, reactive to quick-changing markets towards the increasing customization of products. To establish a sound base for the renewal of manufacturing, this new Leitbild has to be established and diffused throughout the community of industrial actors.

References

Auer, P. and Riegler, C. (1990) *Post-Taylorism: The Enterprise as a Place of Learning Organisational Change*, Stockholm/Berlin: Swedish Work Environment Fund/ Wissenschaftszentrum für Sozialforschung Berlin.

Brandt, D. (1991) *Advanced Experiences with APS Concepts; Design strategies, Experiences — 30 European Case Studies*, FAST occasional paper 246, Brussels: Commission of the European Communities.

Brödner, P. (1985) *Fabrik 2000; Alternative Entwicklungspfade in die Zukunft der Fabrik*, Berlin: Edition Sigma.

Campbell, A., Sorge, A. and Warner, M. (1989) *Microelectronic Product Applications in Great Britain and West-Germany*, Aldershot: Avebury.

Cooley, M., d'Iribarne, A., Martin, T., Ranta, J. and Wobbe, W. (1990) *European Competitiveness in the 21st Century; Integration of Work, Culture and Technology*, Brussels: Commission of the European Communities.

Deiß, M., Döhl, V. and Sauer, D. (1990) *Technikherstellung und Technikanwendung im Werkzeugmaschinenbau*, Frankfurt/New York: Campus Verlag.

Gjerding, A.N., Johnson, B., Kallhauge, L., Laudvall, B.A. and Madson, T. (1990) *Den Forsvundne Produktivitet; Industriel Udvikling i firsernes Danmark*, Charlottenlund: Okonomforbundets Forlag.

Haywood, B. and Bessant, J. (1990) Organisation and integration of production systems, in: Warner, M., Wobbe, W. and Brödner, P. (Eds), *New Technology and Manufacturing Management; Strategic Choices for Flexible Production Systems*, Chichester: Wiley, pp. 75–85.

Hirsch-Kreinsen, H., Schultz-Wild,R., Köhler, C. and von Behr, M. (1990) *Einstieg in die rechnerintegrierte Produktion*, Frankfurt/New York: Campus Verlag.

Kageyama, K. (1993) R&D and productivity in Japanese corporate groups, in: Wobbe, W. (Ed.) with assistance of M. Nakashima, *The Future of Industry in the Global Context; Volume III — Management & Manufacturing*, FAST occasional paper 357, Brussels: Commission of the European Communities, pp. 97–102.

Kern, H. and Schumann, M. (1984) *Das Ende der Arbeitsteilung?; Rationalisierung in der industriellen Produktion*, München: Beck.

Kidd, P.T. (1990) *Organisation, People, and Technology in European Manufacturing*, FAST occasional paper 247, Brussels: Commission of the European Communities.

Lane, C. (1989) *Management and Labour in Europe; The Industrial Enterprise in Germany, Britain and France*, Aldershot: Edward Elgar.

Lehner, F. (1991) *Antropocentric Production Systems: The European Response to Advanced Manufacturing and Globalisation*, FAST occasional paper 248, Brussels: Commission of the European Communities.

Rauner, F. and Ruth, K. (1991) *The Prospects of Anthropocentric Production Systems — A World Comparison of Production Models*, FAST occasional paper 249, Brussels: Commission of the European Communities.

Wobbe, W. (1991) *Anthropocentric Production Systems: A Strategic Issue for Europe*, FAST occasional paper 245, Brussels: Commission of the European Communities.

Womack, J.P., Jones, D.T. and Roos, D. (1990) *The Machine that Changed the World*, New York: Rawson Associates.

3

Modern Sociotechnology: Exploring the Frontiers

Mark van Bijsterveld and Fred Huijgen

Introduction

The classical design principles of *scientific management* and the *bureaucratic model* have resulted in two basic organizational structures: the line structure (every product passes every workstation) and the functional structure (every product passes every functional department) (Kuipers and van Amelsvoort, 1990). Although both structures have major deficiencies, such as inflexibility, poor quality assurance and lack of innovativeness, they flourished during a period characterized by continually growing market demands and the large supply of low skilled workers. These structures were most efficient in the situation of mass production. During the last two decades, however, the organizational environment has changed rapidly. Organizations find themselves confronted with changing market demands, such as shorter and more reliable delivery times, a high and constant product quality, a high flexibility of both production quantities and variants, shortening of product innovation time and are forced to become more customer oriented (Verschuur *et al.*, 1989; van Amelsvoort *et al.*, 1991). The new external function demands cannot be met by traditional organizational structures, so new ways have to be found. Some organizations are trying to find the solution by improving or automating parts of their overall organizational system (e.g. the planning department or assembly line). Others try to find a more integral solution and develop or adopt an entirely new organizational structure.

Kern and Schumann (1984) noticed within some of Germany's main industries (machine building, chemical and automotive industries) a development towards a new way of structuring the organization, which they typified as *Neues Produktionskonzept* (New Production Concept). Within these industries, a renewed craftsmanship was reintroduced which had distinct effects on both the organization's internal and external labour market. One of the main characteristics of this new production concept was the attention given to social and organizational aspects along with the technical components of the production system. On other fronts, practitioners and researchers also developed production systems which were fit to meet the changing external function demands and focused on the interdependency of the machinery and the people that work around them.

Symbiotic design approaches aim to analyse and design the technical, social and organizational aspects simultaneously. Researchers in Europe, Australia and the USA are working on the further development of one group of symbiotic design approaches: the modern sociotechnical systems design. In The Netherlands a special variant was developed under the name of 'modern sociotechnology' (MST). This design approach, which aims at creating an organizational structure that is both economically productive and rewarding for the workers, can be seen as a dramatic break with the classical design principles. Also, in other places of the world, production concepts are developed in order to meet the new market demands and improve a company's competitive strength. However, these concepts cannot be described as symbiotic because they have not really abandoned the line and functional organization structure; what they have done is proved themselves in practice by realizing economically successful production structures. One of the challenges of the symbiotic design approaches is that, because they are not mainstream approaches but have been developed between the borders of sciences such as sociology and engineering, they have to prove themselves in practice by designing production systems that can meet or surpass the performance of more traditional approaches (see Latniak, Chapter 5).

Central questions

The central subject of this chapter is the symbiotic design approach of modern sociotechnology. The following two questions are addressed:

- What are the main characteristics of the concept of modern sociotechnology?
- What are the claims of modern sociotechnology regarding its applicability and performance in different cultural and institutional environments and production situations, and are these claims realized in practice?

The chapter starts by presenting the conceptual framework to be used to describe and evaluate the concept of MST. After that, the main elements and characteristics of MST are described, focusing on its origins and goals, design principles and steps, its realized economic and social performance, and the possibilities for transfer. Then, focus is placed on a currently successful and popular production concept called 'lean production' (LP) and its impact in practice. Based on the characteristics of MST and the discussion of the LP concept the weaknesses of the MST concept are then discussed. The chapter ends with some conclusions about the characteristics and performance of MST and on the possibilities for integrating the concepts of LP and MST.

Conceptual framework

The technical and social system

Within the overall organizational system a distinction can be made between what is called a technical system and a social system (see Figure 3.1).

The technical system includes two main dimensions: division of labour and the production situation. The division of labour can vary according to the production technology, the production organization and the labour organization. The production technology comprises that part of the total amount of labour which is carried out by, or with the help of, automatic devices. In other words, the production technology concerns the division of labour between men and machines. The production organization is the result of a specific way of grouping and coupling the performing functions (production,

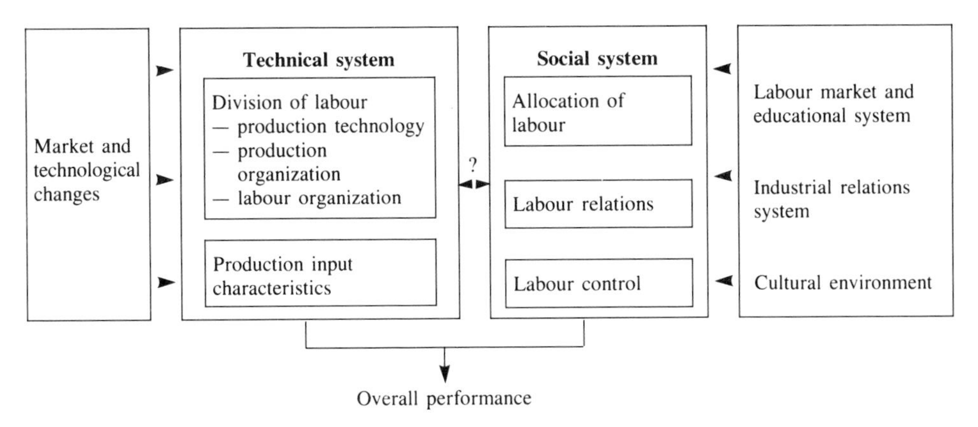

Figure 3.1 The technical and social system

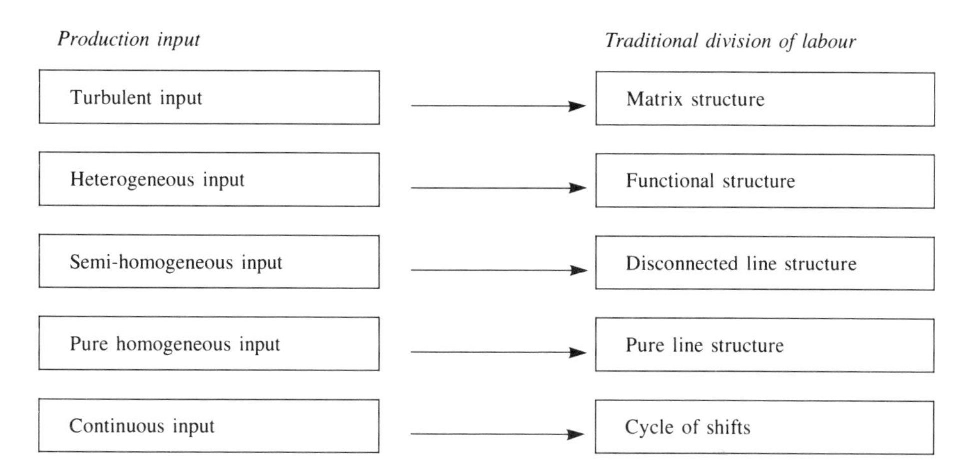

Figure 3.2 The relationship between production input and traditional division of labour structure. Source: Kuipers and van Amelsvoort, 1990, p.100. Reprinted by permission of Kluwer Bedrijfswetenschappen

preparatory, supporting and managerial functions). The labour organization concerns the division of performing functions on the level of individual jobs and/or work teams. The second dimension refers to the characteristics of the production input. According to Kuipers and van Amelsvoort (1990) every type of production input asks for its own structure of the technical system (see Figure 3.2). With respect to the external environment, it can be concluded that the technical system is strongly connected with the characteristics of, and changes in, the product market and the general technological development.

The social system encompasses three dimensions: allocation of labour, labour relations and labour control. The allocation of labour concept refers to the way in which jobs are divided between workers. This concept entails the strategies and activities of recruitment, selection, training and internal mobility. The allocation strategies are influenced by the situation on the relevant labour market and characteristics of the educational system. The concept of labour relations refers to the possibilities for workers to negotiate with management with respect to performance criteria, terms of employment and participation in decision-making. The internal labour relations are influenced by the system of

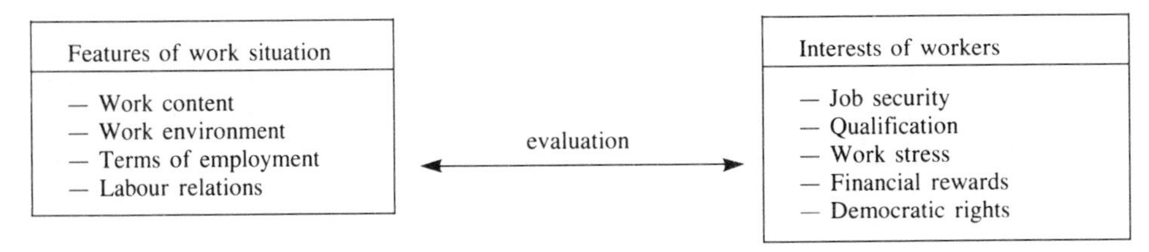

Figure 3.3 The quality of working life

industrial relations. The last dimension within the social system, labour control, refers to the strategies of management with regard to directing the behaviour of workers in order to realize the organizational goals (motivation and socialization strategies). These strategies are influenced by the cultural environment (vocational-, branch- and nation-specific culture).

To what extent the technical and social system interact is difficult to establish. Recent empirical research (ten Have, 1993) demonstrates that the structure of both systems display little association and mutual adaptation.

The quality of work and organization

The overall performance of an organization is a result of the combined operation of the technical and social system. Here we make a distinction between the performance of an organization from an economic point of view and from a social point of view. The economic performance or, as is referred to here, the quality of organization is determined by the following aspects (Peeters, 1993):

- Productivity
- Logistical control
- Flexibility
- Innovative competence
- Quality assurance

The social performance or, as is referred to here, the quality of working life (QWL), is determined by evaluating the work situation in terms of the interests of workers (see Figure 3.3).

With regard to the work situation, four features are distinguished:

1. Work content is determined by the structure of the technical system.
2. The work environment concerns the physical working conditions (excluding stress).
3. The terms of employment concern payment, working times, etc.
4. The labour relations (described earlier) mainly concern the possibilities for negotiating the terms of employment and the participation in the process of decision making.

These four dimensions of the working situation are evaluated on the basis of the workers' interests. Here the following interests are distinguished: job security, the possibilities for realising and improving one's qualifications, the possibilities to influence one's work pace, the financial rewards for a job and the right to express one's opinions and/or complaints during the process of decision making.

Before discussing the MST concept, some additional comments are made with respect to this definition of QWL. First, it should be realized that, even within The Netherlands, a broad variety of descriptions regarding QWL exists and is being used by researchers. Heming (1992) makes a distinction between two groups of descriptions. One group focuses on the structural characteristics of the organization and determines the QWL in an objective way. The QWL is thereby seen as a characteristic of the job. The other group considers QWL as a characteristic of the relationship between individuals and their jobs, and, therefore, the QWL has a more subjective character. The definition given in this chapter falls into the first group and focuses on the more general and 'objective' aspects of working life. Second, it is important to realize that the given definition is influenced by the norms and values that are dominant in Dutch society. It is likely, therefore, that the definitions used in The Netherlands differ from those used in other cultural and institutional settings. Monden (1983), for example, states that 'respect for humanity' is one of the central goals of lean production (LP) in Japanese companies. Although this Japanese concept seems comparable with the presented Dutch concept of quality of working life, it differs in many aspects. QWL focuses on work content, work environment, terms of employment and labour relations, whereas 'respect for humanity' is mainly associated with 'worker-friendly factories' (mainly restricted to improvements in the work environment and ergonomical improvements) and the opportunities for workers to run and improve their own workshops (most likely inspired by economic, and not ethical, motives). Hence, it is questionable whether both descriptions are comparable. Another question that immediately comes to mind is whether it is acceptable to evaluate a Japanese production concept, such as LP, using a Dutch definition of the quality of working life. The intention here is not to answer this question, but simply to raise it, so that the reader may be aware of this dilemma.

Having diverted slightly from the original path of this chapter, the focus will now return to the characteristics of MST.

Modern sociotechnology

Origin and goal

Sociotechnical systems design (STSD) is an integral design method, that is, 'an applied science which aims at improving the quality of work and organization through adaptation or fundamental redesign of contents and composition of technology and human tasks' (van Eijnatten, 1993: 9). This integral design method is over 40 years old, and since its inception three development trajectories can be distinguished (van Eijnatten, 1993):

1. (1949–1959) The period of the sociotechnical pioneering work.
2. (1959–1971) The period of classical STSD.
3. (1971–) The period of modern STSD.

In this chapter, the concentration is on the period of modern STSD. During this period four variants were developed in different countries on different continents (van Eijnatten, 1993). The Australian variant of modern STSD 'participative design' tries to stimulate workers to complete the sociotechnical design process themselves. The cradle of another variant called 'democratic dialogue' was in the Scandinavian countries. This tries to create a successful dialogue between the different participants in the organization. The North-American variant is basically a continuation of the ideas of the classical STSD and heavily relies on the concept of semi-autonomous work groups or so-called 'empowered teams'. The Dutch variant, MST, will now be examined more thoroughly.

In The Netherlands the classical sociotechnical field experiments were conducted by van Beinum and his co-workers (1967) in a cheque-clearing organization (Kuipers and van Amelsvoort, 1990). During this classical period, sociotechnical field experiments were based mainly on common sense and the experience of a few sociotechnical specialists (de Sitter, 1989). However, a coherent and detailed theoretical model was required to spread and apply the sociotechnical ideas on a larger scale. During the 1970s and 1980s, de Sitter and his colleagues at the Eindhoven University of Technology developed the Dutch variant of modern STSD: modern sociotechnology. The MST concept was based on practical experiences and a firm theoretical base: the sociotechnical systems theory. De Sitter's MST concept was later enlarged with a model of change by means of workers' participation after training, and was relabelled 'integral organizational renewal' (van Eijnatten, 1993).

MST is a normative design method based on general design principles and is applicable in a variety of settings and organizations. It is developed to design an organizational structure that can meet external function demands. According to de Sitter (1991) these demands are:

- Flexibility: the possibilities to alter the production process effectively and efficiently.
- Controllability: the use of these possibilities.
- Quality of work and of industrial relations: the capacity to control one's own work and possibility of participating in the decision-making process.

In order to meet these demands MST aims at simultaneously maximizing the quality of organization (flexibility and controllability of the production process) and the quality of working life. For the improvement of an organization's overall performance it is essential not to focus on one of these aspects, either quality of organization or of working life, but to use an integral approach.

Modern sociotechnology is a design method for the structure of the technical system. It thereby makes a distinction between the structures of three interrelated aspect-systems (de Sitter and den Hertog, 1990; Bemelmans, 1991; van Eijnatten, 1993):

- Production structure: the grouping and coupling of performance functions.
- Control structure: the allocation, selection and coupling of control cycles.
- Information structure: 'in sociotechnical design, an information structure is derived from the production and control structure and design questions are concerned mainly with technical matters related to sensing, coding, retrieval and transfer of data' (van Eijnatten, 1993: 179).

Based on the sociotechnical systems theory, de Sitter developed three basic design principles and four design steps in order to redesign the organizational structure.

Sociotechnical design principles

MST is founded on three basic design principles for sociotechnical redesign (de Sitter and den Hertog, 1990; de Sitter, 1991).

Principle 1: The law of requisite variety

This principle, introduced by Ashby (1969), states that the internal diversity of self-organizing systems should match the variation within its relevant environment. System controllability can be seen as a function of the balance between the available internal possibilities for process variation and the required process variation. Control, here, refers to the shaping of structural conditions for opportunities to formulate and realize specific goals.

$$\text{System controllability} = \frac{\text{possibilities for process variation}}{\text{required process variation}}$$

A system is effective and efficient when there is a perfect match between possibilities and requirements. Too many available possibilities create an inefficient system and insufficient possibilities result in an ineffective system.

Principle 2: Reduction of structural complexity increases controllability

A (social) system consists of elements and relations. Its complexity, therefore, can be seen as a function of (de Sitter, 1991):

- the number of different elements;
- the number of different relations;
- the stability of the elements and relations in time.

The greater a system's complexity the harder it is to control its processes, and the greater the chance that disturbances occur. 'The basic principle of integrative design should therefore be to reduce disturbance probabilities by a reduction of impending variety (*authors' note*: variety here implies external and internal variety) and to reduce disturbance sensitivity by an increase in control capacity' (de Sitter and den Hertog, 1990: 16).

Principle 3: The design of the structure precedes that of process technology

The effective and efficient utilization of technology depends upon the architecture of the system structure in which it is applied, because the structure determines how machines and instruments are coupled to input elements and output receivers (de Sitter and den Hertog, 1990). Also, de Sitter (1991) states that the technical requirements can therefore only be determined after the structure of the optimal integrated system has been designed. The adaptation of the structure to the technology is only permitted if it appears to be impossible to meet the technical requirement of the integrated structure design.

Sociotechnical design steps

In order to reach the objectives stated, that is: minimum complexity; minimum disturbance sensitivity and maximum controllability; flexibility and quality of working life, four design steps have to be taken. These steps are aimed at redesigning specific structural parameters and should be taken subsequently.

1. Parallelization

In this first step, in order to reduce the required variation and keep it as small as possible, external variation is diminished by splitting the production process into several

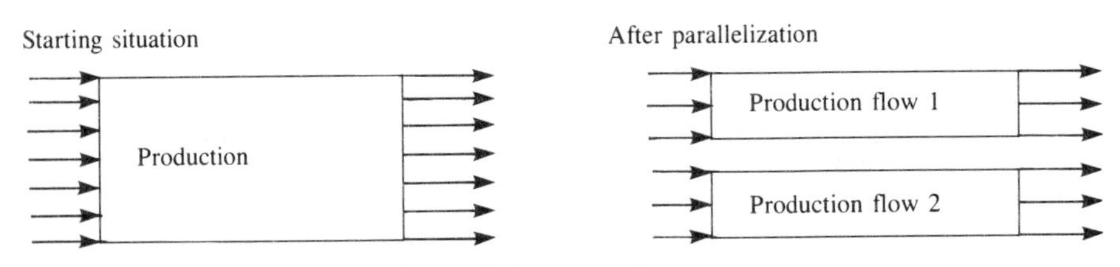

Figure 3.4 Parallellization

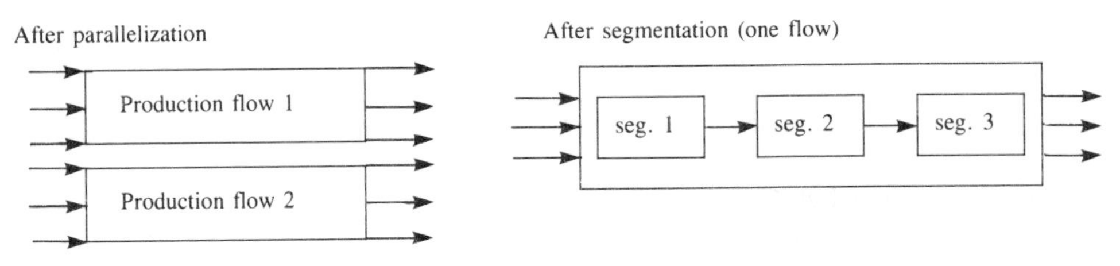

Figure 3.5 Segmentation

production flows, as shown in Figure 3.4. Instead of grouping and coupling *activities* of the same kind, *orders* of the same kind are grouped and coupled (de Sitter, 1991). The mutual characteristic of these orders can be their market segment (e.g. bicycles for Europe or the USA) or their product/service characteristics (e.g. arm-chairs or lounge suites). The effect of parallelization is a significant reduction of input complexity, because the maximum input variation per production flow is substantially smaller than that met originally by the production process.

2. Segmentation

Having created independent product flows, the next step focuses on the reduction in the internal variation within the production flows. The internal variation is caused by the number of interfaces between performance functions. A reduction in interfaces, therefore, will result in a reduction of internal variation and complexity (de Sitter and den Hertog, 1990). This reduction is realized by clustering the performance functions into segments. The segments are created in such a way that the number of the internal interfaces between functions is maximized and the number of interfaces between segments is minimized (see Figure 3.5). The following design criteria are used for creating independent production segments (de Sitter, 1991):

- The activities within a segment are mutually dependent with respect to the sequence of manufacturing and balancing of the work load.
- The activities within a segment are directly dependent on each other with respect to the controllability of quality.
- The activities within a segment should have a high amount of mutual input and output relations.

3. Formation of whole task groups

The first two design steps have realized a reduction in the required process variation. The next step focuses on increasing the available possibilities for process variation by

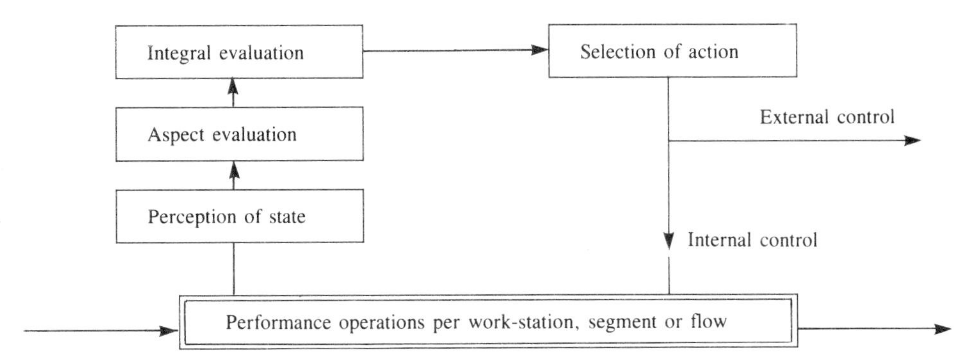

Figure 3.6 The control cycle. Source: de Sitter, 1989, p.242. Reprinted by permission of L.U. de Sitter

designing the internal structure of the segments. To realize this increase, task groups (either whole or semi-autonomous) are formed at the lowest level of the production structure. They are responsible for producing a complete product or service (which can be a component or part of a larger product/service) and can be characterized by the following ingredients: strong internal cohesion, multi-skilled tasks, multi-functional workers, coordination possibilities, shared responsibilities, participation in the coordination process between segments, complete internal regulating of the process, and democratic dialogues (de Sitter and den Hertog, 1990; de Sitter, 1991; van Eijnatten, 1993). Ideally, the members of the whole task groups are responsible for, and capable of, doing all the occurring tasks and activities in the segment, including supportive and preparatory activities. In practice, however, the range of skills mastered by a worker and the range of tasks done depends upon his or her potential and personal ambition (Kuipers and van Amelsvoort, 1990).

4. The control structure

A new control structure is required to make optimal use of the internal possibilities for a flexible production process and to improve the controllability of this process. The complexity of controlling the production process can be decreased by diminishing the amount of required control information and increasing the available control information. This reduction in complexity is realized partially by diminishing the variation within, and the complexity of, the production structure; but more improvements can be made. After the top-down redesign process of the production structure, the redesign of the control structure will be carried out bottom-up, starting synchronously with the formation of the whole-task groups in step 3. 'The control cycle (see Figure 3.6) is the building stone of control structures. In its elementary form the control cycle consists of four interrelated functions' (de Sitter and den Hertog, 1990: 19):

- perception of process states;
- evaluation per aspect;
- integral evaluation of aspects;
- selection of the appropriate control action.

The effectiveness of the control structure is dependent on the reliability, actuality, completeness and relevance of the available control information (which itself is a function of the separation of performance and control functions); control specialization (e.g.

Table 3.1 Weaknesses of the traditional division of labour structure

Economic weaknesses	Social weaknesses
Lack of flexibility	High turnover rates
Poor use of capacity	High absenteeism
Low productivity	Bad social atmosphere
Lack of quality control	High work stress
Generally too high costs	Signs of dissatisfaction

Source: de Sitter, 1989, p.229. Reprinted by permission of L.U. de Sitter

product quality, personnel); control differentiation (operational, structural and strategic); and the division of control functions (perception, evaluation and selection). The main design principle for the control structure, therefore, is to assign as many (complete) control functions as effectively and efficiently possible to the task group. Only those control tasks extending over two or more segments are assigned to a control cycle at a higher level.

The performance of 'sociotechnical companies'

From empirical research done by Warnecke (1977) it became clear that the traditional structure of the division of labour fell short, both of the economic and the social performance (see Table 3.1). Therefore new structures for the division of labour, such as modern sociotechnology, had to be found.

Advocates of MST do not explicitly claim the universal applicability of their design principles and steps. However, their implicit claim is that MST is the best integral concept available for organizations that are confronted with external function demands for flexibility, controllability and quality of working life. In discussing the performance of companies that have adopted the MST concept, three aspects are emphasized: their economic and social performance in different production situations, and the possibilities for transferring the MST concept to companies in a different cultural and institutional setting.

Economic performance

One of the central goals of MST is to maximize the economic performance or quality of organization of a company. The question here is whether it succeeds? It is often said that MST will at its best realize an unchanged economic performance in a company that has dramatically improved its QWL. The slow death of the 8-year-long sociotechnical project 'Quality of working life and organization' within the Dutch truck producer, DAF, was one of the events that strengthens the position of critics. It is questioned whether the concept of MST can (economically speaking) compete with other more traditional production concepts. Therefore, the improvements realized in a variety of organizations adopting the MST concept are highlighted here.

Peeters (1993) evaluated the results of the introduction of whole-task groups within four companies in the clothing industry, all structured to manufacture small series. The best results were gained for logistical control (decreasing throughput time and increasing delivery reliability) and innovation (an increase in process and product improvements). Van Amelsvoort *et al.* (1991) conclude, after implementing MST at a Philips plant, that the main economic improvements were realized in the controllability of the production process, the reduction of the throughput time and the flexibility of the production process and workforce. An evaluation of the introduction of the MST concept in the tobacco factory of Van Nelle by Casparie (1989) revealed the following results: the increase in productivity excelled the norms established in advance and the flexibility of the production process improved clearly thanks to the use of information technology by the whole task groups. Boonstra (1991) introduced the concept of MST in the Dutch subsidiary company of a large international insurance company. The process of change, which took about 2 years, realized a sociotechnical structure in which the efficiency and effectiveness (reduction of throughput time) of the organization and of the quality of its services improved substantially. Also, Terra (1988) reports an improvement in productivity and quality due to the adoption of MST. In this company, which is part of the packing industry, the improvements in productivity must, however, be ascribed mainly to the decrease in absences due to illness (probably because of the improved quality of working life). We also know that the concept of MST is adopted in the automotive industry by companies such as DAF and Nedcar (see Verschuur *et al.*, 1989; Aertsen and Benders, 1993). Unfortunately, little concrete data are available in respect of the economic performance of these companies after they (partially) adopted MST.

The general conclusion, based on these case studies, should be that sociotechnical companies do show an improvement in economic performance and seem to succeed, up to a certain level, in meeting the dominating external function demands of flexibility, quality assurance, improved productivity and high controllability of the production process. The companies and industries in which MST, from an economic point of view, was most successfully implemented had originally in most cases a functional structure (heterogeneous production input) and a disconnected line structure (semi-homogeneous production input).

Social performance

MST has, in contrast to other production concepts, the quality of working life as a central goal. The design of a humanistic, democratic and unstressful work organization is not seen as a separate design step within the MST concept, but is an integral part of the four design steps described earlier. In discussing the anticipated and unanticipated effects on QWL that can be found in sociotechnical companies, the reader has to bear in mind that MST only provides explicit principles for designing the work content.

Work content

Creating motivating and fulfilling jobs which give workers a chance to develop themselves, and in which their skill and knowledge are optimally used, is one of the main goals and achievements of MST. Creating a company with a high QWL is not just a way of improving a company's economic performance, but is seen as an end in itself. High-quality jobs designed according to the principles of MST are broad (multi-skilled). They include production, control and preparatory activities and provide workers with some autonomy and control over their work pace and work routines (de Sitter, 1991; Kuipers and van Amelsvoort, 1990; van Hootegem and Janssens, 1993).

MST's aim to achieve a high quality of work can be criticized for neglecting the fact that all workers are individuals and have different needs and expectations of their work (Heming, 1992). It can be claimed, therefore, that not all workers want challenging and autonomous jobs. If this claim is endorsed by companies adopting MST, it could (and maybe it should?) influence their selection and training programme.

Work environment

Although the design method of MST does not give explicit guidelines for designing a work environment, in practice we find (as was expected) a work environment that has been created to maximally support the functioning of workers (Kuipers and van Amelsvoort, 1990). First, the work place is safe, healthy and clean (e.g. lack of noise, a lot of work space); this is partly the result of the strict legal criteria for companies. Second, machines are adapted to the designed tasks and to the specific requirements of each worker (e.g. ergonomic work tools).

Terms of employment

The reward system of sociotechnical companies has a somewhat dualistic character. The system has as its goal to reward workers for their efforts and to motivate them to achieve maximum performance and quality. This requirement demands a highly flexible system of financial rewards, paying each group and group member according to performance (van Amelsvoort and Scholtes, 1994). However, this practice is tempered by the existing labour relations system, social norms and laws, all demanding a high degree of financial security for workers. In practice, often the larger part of the salaries are fixed, based on skills and knowledge, and only a small part is flexible, existing of bonuses based on group performance (Kuipers and van Amelsvoort, 1990).

Labour relations

In practice 'sociotechnical companies' try to create a democratic organizational culture, supported by a decision-making structure through which workers can ask questions, make critical remarks and give suggestions for improvements. Another characteristic of a democratic organization is the participation of workers or some representatives in those committees discussing and making decisions about the companies short and long term strategies (Rehder, 1992).

Possibilities for transfer

It can be questioned whether or not it is possible to transfer the concept of MST to companies in countries with a different institutional and cultural background: is this possible or is this concept too strongly influenced by its cultural background? Here we pose that the concept of MST is strongly influenced by its Dutch (or more generally speaking European) cultural and institutional background and that this indeed may cause problems for transfer. The emphasis that MST lays on maximizing the QWL by structural redesign can be seen as a result of a wealthy society with high social standards, in which individuals are nurtured and protected. Other countries, with different environments and background, may have adopted a quite different attitude towards individuals and their psychological and social needs. In the USA, an attitude of individual wealth and welfare dominates. Each individual has to employ activities to ensure and improve his or her own situation. In Japan, the emphasis is laid on the economic achievements and well-being of groups instead of individuals. These differences legitimate the question whether MST can be fully transferred to areas with a different cultural and institutional environment. Especially the idea of semi-autonomous groups, giving a great amount of autonomy and initiative to workers, may be difficult for countries where people are used to the fact that others standardize their work and tell them what to do when problems occur. However this question cannot be answered sufficiently because the MST concept has not (on a reasonable scale) been adopted by companies outside Europe. The new Saturn plant of General Motors is designed, with considerable social and reasonable economic results, according to a production concept that seems to embrace some sociotechnical principles (van Hootegem and Janssens, 1993), but further studies and initiatives have to prove whether it is possible to successfully adopt MST in other institutional and cultural settings.

Lean production

In the above the main characteristics of MST and the economic and social performance of companies adopting this production concept have been introduced and described. Of course it is not the case that MST or other symbiotic design approaches are the only production concepts that are developed to meet the new and changing external function demands with which organizations find themselves confronted. One of these other, highly successful, production concepts is lean production. Here this concept and the performance of companies who have adopted it will be described briefly, so that it may be compared with the concept of MST.

The concept of lean production

The term 'lean production' (LP) was introduced by Krafcik (1988) as part of the International Motor Vehicle Program (IMVP), an extensive comparative study into the automotive industry lead by researchers from MIT. LP became well known as a production concept after the publication of the book *The Machine that Changed the World* (Womack *et al.*, 1990), but can be seen as merely a new label for the already well-known *Toyota Production System* (see Monden, 1983; Toyota, 1992). The basic goal of the LP concept is to maximize a company's economic performance. This maximalization can be accomplished by focusing on four sub-goals (Monden, 1983):

- Cost reduction: by eliminating waste.
- Quality assurance: assures that each process will supply only good units to next processes.
- Quantity control or flexible production: enables the system to adapt to daily and monthly fluctuations in demand.
- Respect for humanity: must be cultivated while the system utilizes the human resources to attain its cost objectives.

In order to realize this maximization Japanese companies make use of a wide variety of manufacturing techniques and personnel practices. Some of these techniques and practices are: Just-in-time (JIT) manufacturing and supply, total quality control, *shoijnka*, *kaizen*, life-time employment, on-the-job training, (bi)annual appraisal system (see Monden, 1983; Womack *et al.*, 1990; Mayes and Ogiwara, 1992; Young, 1992; Benders and Aertsen, 1993; van Hootegem and Janssens, 1993). The characteristics of LP that seem to account for its success are: its manufacturing techniques and personnel practices; the intensive interdependence with its suppliers and its purchasers (compare with the concept of value stream; Womack and Jones, 1994), and the manufacturability of its products (Womack *et al.*, 1990; Young, 1992; van Hootegem and Janssens, 1993). None of these characteristics are exclusive to LP, however, rather a number of well-known and new strategies, techniques and practices are adopted, refined and formed into a coherent concept (Benders and Aertsen, 1993). LP is, in its traditional Japanese form, more than just a set of loosely coupled manufacturing techniques and personnel practices. Alders (1993) concludes that the strength of the LP concept lies in the consistency and harmony of its manufacturing techniques and personnel practices. Although each element can be introduced individually (see Oliver and Wilkinson, 1989), the strength of the concept lies in its application as a whole.

Womack *et al.* (1990) presented data which clearly show the superior economic performance of lean companies in the automotive industry and made the concept of LP very popular. Other studies (Oliver and Wilkinson, 1989; Mayes and Ogiwara, 1992; van Hootegem and Janssens, 1993) confirm these findings. The success of the LP concept has to be ascribed to the spectacular improvements in productivity, quality assurance and logistical control, and, to a lesser extent, the flexibility and innovativeness it realized in companies across several industries (automotive industry, the industry for manufacturing components and the consumer electronics industry). These industries can, according to Young (1992), be characterized by their repetitive manufacturing system (pure line structure), that is they have a medium to high batch size and few product types (pure homogenous production input). The repetitive manufacturing companies who have successfully adopted lean manufacturing techniques and personnel practices can be found in a wide variety of cultural and institutional settings, adapting some of the manufacturing techniques and (most of the) personnel practices to their local circumstances. The adaptation to the local situation seems to have only a minor tempering effect on the improvement of their economic performance (Oliver and Wilkinson, 1989; Mayes and Ogiwara, 1992; Dhondt and Pot, 1993; van Hootegem and Janssens, 1993). The claim of Womack, that the principles of LP are universal and 'can be applied in every industry across the globe' (Womack *et al.*, 1990: 8), has to be tempered. The principles that can, and have, successfully been transferred to other institutional and cultural settings are mainly the lean manufacturing techniques, and then mainly to companies that have a homogenous production input.

Although respect for humanity is one of the goals of LP, the quality of working life in many lean companies can be described as unsatisfactory (according to European standards). Workers have to function continually at their maximum, their jobs are strongly standardized, the team culture is one of strong social control and peer pressure, and they are uncertain of the rewards they receive for their efforts (Parker and Slaughter, 1988; Rehder, 1992). Other less fundamental bright spots are the rotation system, lifetime employment (albeit only in large Japanese companies), team responsibility for quality and supportive tasks and current improvements of the work environment (van Hootegem and Janssens, 1993).

The impact of lean production

The main impact of the IMVP is thought to be not the presentation of Japanese or lean manufacturing techniques and personnel practices, but the presentation of the data that show the superior economic performance of these techniques and practices. These production data have become the benchmark for many companies using traditional or symbiotic (e.g. sociotechnical) production structures. In a recent discussion with a manager of Nedcar (a joint venture of Volvo with Mitsubishi in The Netherlands) this conclusion was confirmed. He said that the people in his organization were anxiously waiting on the new and delayed IMVP data, because they wanted to know whether the current restructuring of the Nedcar organization was already paying off and improving their competitive strength. This example is very characteristic of the use of research data and production concepts in practice. Research data like those of the IMVP are giving companies a sense in which direction they have to steer their strategy and (long-term) goals. Production concepts are in practice mostly seen as means for realizing the formulated strategy and goals. What we see in practice is that (parts of) several production concepts are used indifferently at the same time, probably because of one of the following reasons. First, companies use production concepts in a very pragmatic way, they have to bring the company closer to its goal — equalling or surpassing the performance of excellent competitors. It then does not matter. To quote the manager of Nedcar, 'whether what we do is lean or sociotechnical is unimportant, as long as it works for us'. The second reason is a far less rational one. Companies often have a very limited understanding of the principles and content of a certain production concept (Schumann *et al.*, 1994). In practice we see then, as a result of this lack of understanding, that companies use techniques and practices of several new and traditional concepts indifferently under the label of, for example, LP or MST. A good example here is the case of the Dutch truck producer DAF (Aertsen and Benders, 1993). After the stagnation and gradual withdrawal of the sociotechnical 'Quality of working life and organization' programme, DAF (or rather the DAF board of directors) got enthusiastic about LP. The lack of understanding of the LP concept and of an overall design philosophy was the main reason that the introduction of LP within DAF did not seem to result in a radical break with its (sociotechnical) past. Aertsen and Benders even conclude that 'it is possible to carry on sociotechnical experiments under the new label of lean production' (1993: 19).

Weaknesses of modern sociotechnology

The goal of this section is to establish the strengths and weaknesses of MST and on which fronts its threats and future opportunities lie. The importance of this analysis lies in the assumption that a static production concept cannot remain successful in a dynamic environment. It is also important for as radically new production concepts as MST, to continually work on its refinement and improvement. Lessons can be learned from developments in several other countries.

First there are the lessons that can be learned from Japan. Earlier it was mentioned that the success of the Japanese production concepts lies in the coherence it creates between manufacturing techniques, personnel practices, structural dependency relations with suppliers and purchasers and the manufacturability of its products (although the latter is not a specific characteristic of Japanese companies and can be found also in other automotive companies (see Womack *et al.*, 1990; van Hootegem and Janssens, 1993)).

The success of MST, on the other hand, depends totally on its design of the technical system. In the original MST concept no explicit attention was paid to the design of the social system and the structuring of an external network. In practice, however, the use of the concept of MST differs somewhat from the theoretical concept. This is inevitable, of course, because one cannot change an organization by designing and implementing a new organizational structure without giving attention to the design of the social system and its external network with suppliers and buyers. Kuipers and van Amelsvoort (1990) describe some of the personnel practices that were developed in the sociotechnical design project they conducted. More explicit design principles need to be developed, however, especially in respect to the external network, and integrated with the design principles for structuring the technical system.

More and other lessons can be learned, however, and the Swedish sociotechnical approach has a strong base in participation, team work and a democratic leadership style (see Karlsson, Chapter 4). The 'Democratic Dialogue Model' (Gustavsen, 1992; Engelstad and Gustavsen, 1993) emphasizes how the design process should be shaped, not how the production process should be designed. In the concept of Democratic Dialogue 'the communication component is explicitly put into the foreground' (van Eijnatten, 1993: 71). Another approach, also focusing on the design process but emphasizing both the communication and the design component, is the Australian concept of 'Participative Design' (Emery, 1989). This method 'enables employees, (middle) management and union representatives to jointly take over the task and organization design from the start of the project' (van Eijnatten, 1993: 147). Although both concepts have different backgrounds and goals for establishing communication between, and participation of, the stakeholders in the design process, both concepts focus on the shaping of the design process and not on the structure that has to be designed. This explicit attention for the design process itself is lacking in the concept of MST. An often heard accusation against (Dutch) sociotechnical practitioners is that they act like experts, advising the organization on how to structure the division of labour without letting the stakeholders participate in the design process and without helping them to get from their current situation into the desired sociotechnical situation (den Hertog and Dankbaar, 1989; Martens, 1994). Whether these critical remarks are entirely true or not is still a question of debate and will not be addressed here.

It is clear that one could describe the debate as a dilemma between, on the one hand, the vision of MST on how the technical system should be structured and, on the other hand, the democratic rights of the stakeholders within and around the organization (e.g. employees, management, unions). In other words, it deals with the question whether participation must be seen as a (moral) goal of the change process, thereby saying that the result of the participative design process is of less importance, or whether participation is primarily instrumental and one of the means to ensure, by using the available knowledge within the organization and eliminating resistance against change, the successful implementation of the sociotechnical organization structure. It has been mentioned already that that is a dilemma, one must chose one of the alternatives. One of the weaknesses of MST is that it does not explicitly make such a choice and, therefore, does not provide any principles for shaping the design process. The 'concept of integral organizational renewal' (van Eijnatten, 1993) has enlarged the MST concept with a model of change. It tries to ensure the successful adoption of the MST concept by letting workers participate in the design process after a training (or should it be called a manipulation?) programme. The goal of this participation is primarily instrumental, aiming at reducing any possible resistance. It can, however, be questioned whether this model of change can be called sociotechnical, because it lacks a sound (social systems) theoretical foundation.

Another aspect of the MST concept that is not well worked out is the role and influence of the technical factors. MST aims at designing an organization in which people are the central elements of the production process. Machines, computers and other physical and mechanical tools are treated as if they were completely flexible and adaptable to the designed organization and tasks. Unfortunately, how to determine the necessary characteristics of the technical factors, based on the designed structure, is left unclear. Van der Zwaan, however, states that 'it is seen that in many cases of mechanization and automation the technological development continues to operate so autonomously that it compels the adaption of people to the technological development rather than in reverse' (1975: 155). In practice, machine and computer are often not as flexible as sociotechnical practitioners would like them to be. The constraints of these artefacts force sociotechnical designers to adjustment of their original and optimal design. Although MST acknowledges that in some cases the organizational structure must be adapted to the technical factors (de Sitter, 1991), it does not provide any design principles for how to deal effectively with these problems. Thus, generally speaking, the MST concept pays little or no attention to the mutual interdependency between structure, tasks and technical factors. That this can become a threat for the concept of MST may become clear when we realize that these factors (e.g. robotics, information technology, etc.) still have a growing and vital influence on the quality of organization and work.

MST has its theoretical base in the socio-technical systems theory. The design principles and steps for structuring the organization are mainly derived from this theory. If MST accepts the above, given critics and warnings, and strives for enlarging and improving the current concept, it is of fundamental importance that these changes have a sound theoretical base. Such a base provides a steady norm from which new principles can be derived and against which new developments can be held for an critical evaluation. This base should preferably be found in the social systems theory or one that is compatible with it.

Discussion and conclusions

In the last part of this chapter an attempt will be made to answer the following questions:

- What are the claims regarding MST's applicability and performance and are these claims realized in practice?
- What is the function of another production concept, LP, for MST and is it possible to integrate these concepts?

Theoretical foundation and reach of modern sociotechnology

During the first stages of the development of the socio-technical systems design practical experiences and field experiments played a major role. Thanks to the theoretical work of de Sitter in particular, a Dutch variant of STSD (modern sociotechnology) was developed on a sound theoretical base: social systems theory. The described design principles and steps are mainly derived from this special variant of the social systems theory. MST means a fundamental rethinking and radical redesign of the overall structure. Instead of a

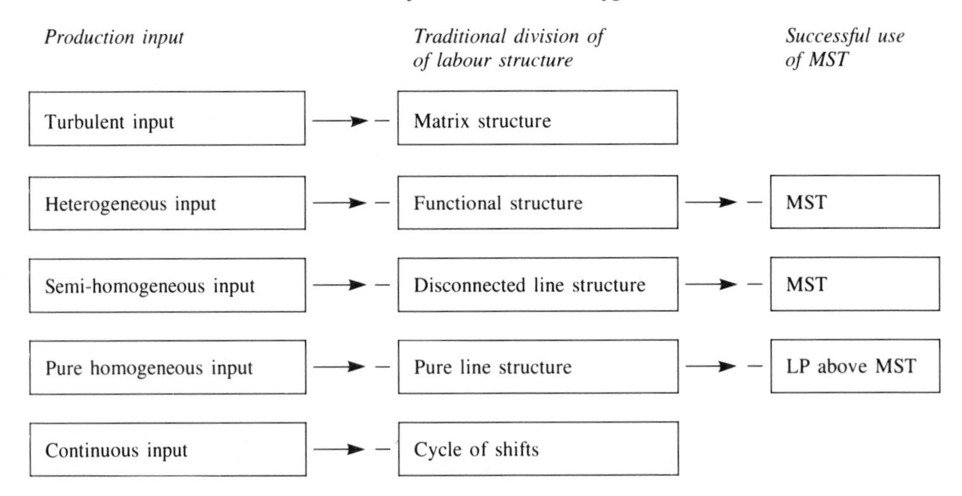

Figure 3.7 Production input and the economic successful use of modern sociotechnology

complex structure with simple tasks, a simple structure with complex tasks is designed. As such, MST aims at the improvement of what could be typified as the chance structure for meaningful human work situations, by designing a division of labour which provides opportunities but without prescribing how they should be used. For it is only the division of labour which is the object of the redesign process, whereas the structure and the functioning of the social system, the relations with other organizations (e.g. suppliers, distributers), and the process of design and change is not at stake. It is desirable that the MST concept is enlarged with well-founded (e.g. socio-technical systems theory) design principles and steps for these new areas.

Pretences and practice

The explicit, or implicit, pretence of MST is that, given the dominance of flexibility, controllability and quality of working life as external function demands, its design principles and steps are universal and can be applied successfully in a wide variety of industries and countries. Practical experience, however, shows a somewhat different picture.

The MST concept is adopted by both industrial (small and large batch size) and service companies in primarily European countries. From the previously mentioned field experiments the conclusion can be drawn that the realized economic improvements by adopting MST are not spectacular, but highest in companies with heterogeneous and semi-homogeneous production input, especially small series or unique products. The main improvements of MST can be found for the aspects' flexibility, innovative competence and logistical control. Although MST can also be used for companies facing a pure homogeneous production input (e.g. the automotive industry), MST does not reach an economic performance that is as high as that of the concept of LP (see Figure 3.7).

By contrast, the improvements for the quality of working life, especially the work content and labour relations, achieved by adopting MST were very high in all of the companies investigated, and substantially higher than in 'lean companies'. The designed semi-autonomous task groups provide workers with jobs that are both fulfilling (they include performing, control and supportive activities), challenging (providing opportunities for developing oneself) and worker-friendly (setting own work pace and working with ergonomic designed machines and instruments). The group and organizational culture in practice can be described as democratic and provides workers with the possibilities to ask questions, complain and give suggestions. From a social point of view we can conclude than modern sociotechnology is in all kinds of (Dutch) organizations more successful than lean production. Whether or not the concept of MST can be successfully transferred to other countries is still a question that lacks a sufficient answer and asks for additional research.

Lean sociotechnology?

One of the most interesting subjects for future research is whether or not it is possible to integrate the concepts of modern sociotechnology and lean production into a single new production concept. The goal of this integration would be to create a sociotechnical concept that can realize higher improvements in the economic performance of organizations without tempering its current success in improving the quality of working life. Without suggesting an answer, some comments can be made about the possibilities and obstacles for such an integration. A clear obstacle is their difference in vision on how to structure the technical system. Where LP can be characterized as a production concept that aims at perfecting the traditional line structure ('the Japanese out-Taylor us all'; Schonberger, 1982: 193), MST means a fundamental rethinking and radical redesign of the technical system. A new and simple organization which is built from complex jobs is designed by MST.

Another obstacle is the fact that MST and LP have been developed in totally different institutional and cultural settings. These settings have strongly influenced the underlying assumptions and norms of the concepts. Earlier it was warned that a concept cannot just be transplanted from one setting to another. The same accounts for the integration of both these, culturally different, concepts. The only fruitful way envisaged for integrating both concepts is by enlarging the concept of MST with some of the manufacturing techniques of LP, e.g. total quality management, kaizen, JIT production and supply. This is possible and desirable because most of the manufacturing techniques are 'neutral' and are a valuable enlargement of the concept of modern sociotechnology.

Acknowledgements

The financial support of the Netherlands Organization for Scientific Research (NWO), the Royal Netherlands Academy of Sciences (KNAW), and the Department of Business Studies of the University of Nijmegen is gratefully acknowledged.

References

Aertsen, F. and Benders, J. (1993) 'Tricks and Trucks: Ten Year of Organizational Renewal at DAF?', research memorandum FEW 581, Tilburg: Tilburg University.

Alders, B. (1993) Lean Production: is er sprake van een Westeuropees perspectief?, *M&O*, **49** (1), 39–54.

van Amelsvoort, P. and Scholtes, G. (1994) *Zelfsturende teams: ontwerpen, invoeren en begeleiden*, Oss: ST-Groep.

van Amelsvoort, P., Knol, G., Leyen, D. and de Weerd, H. (1991) Integrale organisatievernieuwing Philips Stadskanaal, *Bedrijfskunde*, **63** (2), pp. 208–217.

Ashby, W.R. (1969) Self-Regulation and Requisite Variety, in: Emery, F.E. (Ed.), *Systems Thinking; Selected Readings*, Harmondsworth: Penguin Books, pp. 105–124.

van Beinum, H.J.J., van Gils, H. and Verhagen, E.J. (1967) *Taakontwerp en werkorganisatie: een sociotechnisch veldexperiment*, Leiden: NIPG.

Bemelmans, T.A. (1991) *Bestuurlijke informatiesystemen en automatisering*, Deventer/Leiden: Kluwer/Stenfert Kroese.

Benders, J. and Aertsen, F. (1993) Aan de lijn of aan het lijntje: wordt slank produceren de mode?, *Tijdschrift voor Arbeidsvraagstukken*, **9** (3), 263–272.

Boonstra, J.J. (1991) *Integrale organisatie-ontwikkeling*, Culemborg: Lemma.

Casparie, P. (1989) Bij Van Nelle bepalen de taakgroepen zelf wat nodig is, *Arbonieuws*, no. 5.

Dhondt, S. and Pot, F.D. (1993) 'Japanization/lean production in The Netherlands: Consequences for work and economic performance', paper presented at the 10th Annual Conference of the Euro-Asia Management Studies Association, Nürnberg, 18–20 November.

van Eijnatten, F.M. (1993) *The Paradigm that Changed the Work Place*, Assen: Van Gorcum.

Emery, M. (Ed.) (1989) *Participative Design for Participative Democracy*, Canberra: Australian National University, Centre for Continuing Education.

Engelstad, P.H. and Gustavsen, B. (1993) Swedish network development for implementing national work reform strategy, *Human Relations*, **46** (2), 219–248.

Gustavsen, B. (1992) *Dialogue and Development: Theory of Communication, Action Research and the Restructuring of Working Life*, Stockholm: Arbetslivscentrum.

ten Have, K. (1993) *Markt, organisatie en personeel in de industrie*, Tilburg: Tilburg University Press.

Heming, B.H.J. (1992) 'Kwaliteit van arbeid, geautomatiseerd . . .; Een studie naar kwaliteit van arbeid en de relatie tussen automatisering, arbeid en organisatie', PhD dissertation Delft University of Technology.

den Hertog, J.F. and Dankbaar, B. (1989) De sociotechniek bijgesteld, *Gedrag & Organisatie*, **2** (4/5), 269–288.

van Hootegem, G. and Janssens, F. (1993) *Nieuwe arbeidsvormen aan de lopende band: verslag van een field trip naar Saturn, NUMMI en Ford Atlanta*, Leuven: Catholic University of Leuven.

Kern, H. and Schumann, M. (1984) *Das Ende der Arbeitsteilung? Rationalisierung in der industriellen Produktion*, München: Beck.

Krafcik, J.F. (1988) Triumph of the lean production system, *Sloan Management Review*, **30** (1), 41–52.

Kuipers, H. and van Amelsvoort, P. (1990) *Slagvaardig organiseren: een inleiding in the sociotechniek als integrale ontwerpleer*, Deventer: Kluwer Bedrijfswetenschappen.

Martens, W. (1994) 'Arbeidsprocesbenadering en sociotechnische systeemtheorie: hun logica en problemen', unpublished paper University of Nijmegen, Department of Business Studies.

Mayes, D.G. and Ogiwara, Y. (1992) Transplanting Japanese success in the UK, *National Institute Economic Review*, **142**, 99–105.

Monden, Y. (1983) *Toyota Production System*, Norcross: Industrial Engineering and Management Press.

Oliver, N. and Wilkinson, B. (1989) Japanese manufacturing techniques and personnel and industrial relations; practice in Britain: evidence and implications, *British Journal of Industrial Relations*, **27** (1), 73–91.

Parker, M. and Slaughter, J. (1988) Management by stress; behind the scenes at NUMMI Motors, *New York Times*, 4 December, p. 2.

Peeters, M. (1993) Welzijn met dank aan de consument, *M&O*, **49** (5), 371–389.

Rehder, R.R. (1992) Building cars as if people mattered: the Japanese lean system vs. Volvo's uddevalla system, *The Columbia Journal of World Business*, **27** (2), 56–69.

Schonberger, R.J. (1982) *Japanese Manufacturing Techniques; Nine Hidden Lessons in Simplicity*, New York: Free Press.

Schumann M., Baethge-Kinsky, V., Kuhlmann, M., Kurz, C. and Neumann, U. (1994) Zwischen Neuen Produktionskonzepten und Lean Production, *SOFI Mitteilungen*, **21**, 26–35.

de Sitter, L.U. (1989) Moderne sociotechniek, *Gedrag & Organisatie*, **2** (4/5), 222–252.

de Sitter, L.U. (1991) *Cursus Organisatie-analyse en ontwerp*, Module 1-6, Maastricht: MERIT.

de Sitter, L.U. and den Hertog, J.F. (1990) 'Simple Organizations, Complex Jobs: the Dutch Sociotechnical Approach', paper presented at the Annual Conference of the Academy of Management, San Francisco, August 12–15.

Terra, N. (1988) Ruimte voor nieuwe inhoud: reorganiseren met menselijke maat, *Gedrag & Organisatie*, **1** (4), 72–87.

Toyota (1992) *The Toyota Production System*, Toyota City: Toyota.

Verschuur, F.O., Groen, K. and van den Broeck, J. (1989) Een toepassingsgerichte ontwerpstudie voor de vorming van hele taakgroepen bij DAF in Westerlo, België, *Doelmatige Bedrijfsvoering*, **1** (11), 24–30.

Warnecke, H.J. (1977) Neue Arbeitsstrukturen und Materialfluß, *Zeitschrift für industrielle Fertigung*, **67**, 185–190.

Womack, J.P. and Jones, D.T. (1994) From lean production to the lean enterprise, *Harvard Business Review*, **72** (2), 93–103.

Womack, J.P., Jones, D.T. and Roos, D. (1990) *The Machine that Changed the World*, New York: Rawson Associates.

Young, S.M. (1992) A framework for successful adoption and performance of Japanese manufacturing practices in the United States, *Academy of Management Review*, **17** (4), 677–700.

van der Zwaan, A. (1975) The sociotechnical systems approach, *International Journal of Production Research*, **13** (2), 149–163.

4

The Swedish Sociotechnical Approach: Strengths and Weaknesses

Ulf Karlsson

Introduction

In this chapter the Swedish sociotechnical approach is examined and its strengths and weaknesses analysed. During the last two decades Swedish organizational scientists, together with engineers, have been actively involved in the design and development of socially and technically integrated systems. These efforts range from comprehensive in-company design projects to off-site academic experimentation. They have been conducted within the wider context of 'Scandinavian management', which consists of participation, team work and equality, and where formal status is considered less important. The overall goal of Scandinavian management is for everybody to learn more and develop within the job. Within the organization honesty and hard work are valued qualifications while 'rational', rather than 'emotional', thinking within production and administration is taken into consideration when the system is appraised. Against this idea it should be emphasized that Scandinavian management tends to involve 'flat' organizations, which makes it difficult to promote people to another level within the organization while supervisors have to be much more skilled in leadership than those employed by a more traditional organization.

In Sweden, there has been a great deal of design-oriented research carried out, which has been aimed at identifying the best systems for achieving the overall goal of Scandinavian management. The models employed in this research are described and the use of case studies and action-based methods is evaluated, with particular reference to problem definition and conflict resolution. Examples are provided of sociotechnical approaches in different industries and applications in Sweden where, it should be stressed, it has been industrial engineers and not social scientists who have been the main users of sociotechnical ideas.

An important issue in Sweden recently has been the emergence of Japanese management. While many Japanese ideas complement the Swedish sociotechnical approach others, most notably the concept of 'lean production', at first sight appear to be based on different philosophies. It is argued here, however, that certain aspects of the Japanese approach can be incorporated into Swedish management style. In this way broad job design concepts can be devised which can be applied to technology in the future.

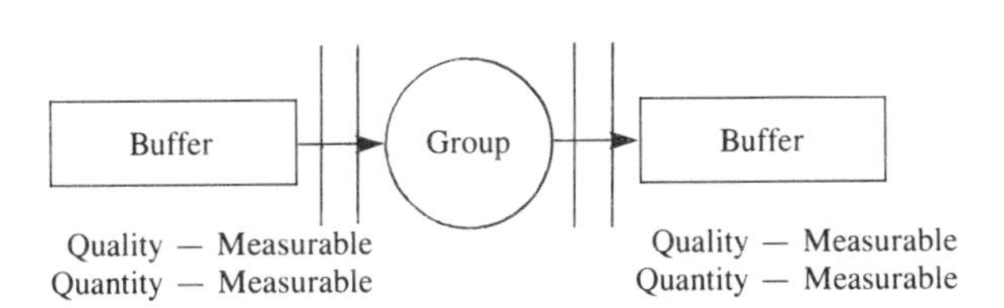

Figure 4.1 Group autonomy

Content of the Swedish sociotechnical approach

Technology analysis and autonomy

In traditional design-oriented research, technology is seen more or less as a determinator that cannot be changed. In the 1970s and 1980s, behavioural scientists tried to design an organization to suit a specific technology like CIM or FMS. Van der Zwaan (1975) pointed out that the sociotechnical approach aims at balancing social systems and technologies to suit each other, and that sociotechnical theory is underdeveloped in the area of analysis of technology and design of technological systems. The reason for this is mainly that the researcher has been a behavioural scientist instead of an engineer. Karlsson (1979) shows that there are methods for designing different technological systems in balance with the social system. However, this requires the methods to analyse old or new technology that Karlsson (1979) has developed. The focus on the technical autonomy and how to achieve this by parallel flow groups, magazines, small buffer stocks and rotations between work stations which lack autonomy are well-known from this study. Engström (1983) has focused upon redesign of the materials flow system and the materials handling system to achieve technical autonomy, but there is still the problem of a lack of methods to analyse and design technology systems in the R&D department.

A further consideration is that change cannot always be achieved. One example is where expensive machines which often restrict flow patterns cannot easily be replaced. Another is where working hours, time and shift work cannot easily be changed.

The autonomy of the groups is based on them being given a certain measure of technical independence in relation to the remainder of production. Group autonomy can be described by the model shown in Figure 4.1.

The implication of Figure 4.1 is that the technical system must ensure:

1. that the group can check how much material is supplied to and delivered by the group;
2. that the group knows when material is to be returned, adjusted or scrapped;
3. that the buffer before and after the group offers opportunity for the group to produce at its own pace.

Autonomy has created the prerequisites for the group to take its own responsibility for certain decisions. The areas where a group can take responsibility are:

1. ensuring that the production programme is followed;
2. checking incoming materials;
3. rectifying defects in incoming materials;
4. return of defective materials;
5. return of materials damaged in transit;
6. transport, including the ordering of transport;

7. maintenance of hand machines and daily inspection of the welding units, jig equipment, flange machines and adhesive pumps cleaning;
8. training of new members;
9. the right to take time off without pay (a maximum of one day);
10. materials requisitions;
11. contact with welding service and jig service in the event of disturbances in production;
12. control of consumables;
13. rationalization work within the department through the suggestions scheme;
14. inspection of their own production and follow-up in succeeding production sections;
15. feedback of defects on random sampling inspection of finished products;
16. adjustments to their own production budget, comprising indirect hours for stand-ins and extra personnel.

The wages consist of a fixed hourly wage with an incentive agreement (usually 1.10 methods-time measurement (MTM)). The group should produce a certain number of products per week. Production over and above this limit will not result in higher wages, and the group need not, therefore, feel the stress often associated with piecework.

A group responsibility allowance is payable according to its increase in responsibility. This allowance is in payment for such duties which cannot be measured by means of the work valuation system of the Swedish Metal Trades Employers' Association/the Swedish Metal Workers' Union. The responsibility which can be delegated to a production group is divided up into five main sections:

1. production responsibility;
2. quality responsibility;
3. economic responsibility;
4. administrative responsibility;
5. social responsibility (no rejection of members).

The acceptance of responsibility in each section can take place in several stages which gradually give an increased allowance. The maximum responsibility allowance payable can amount to 3–5 Sw Kr per hour.

There are rules concerning the manner in which the work within the group is to be carried out. These rules are prepared by the members themselves, together with the shop-floor supervisors. The rules are approved jointly by the works' engineer and the group before being entered into the records kept by each group. If the group considers that a rule does not work in practice, it is free to reformulate it and discuss it with the shop-floor supervisor. The rule is then approved jointly by the group and the works' engineer.

The rules determine the responsibility of the group in relation to other departments and also determine internal matters, such as when and how job rotation, work distribution, etc., can take place.

Change process

A question that is seldom asked is 'who has formulated the problem definition?'. A company has a lot of different departments and they all think they know the nature of the problem. Those that often define the problem are the senior management group and in

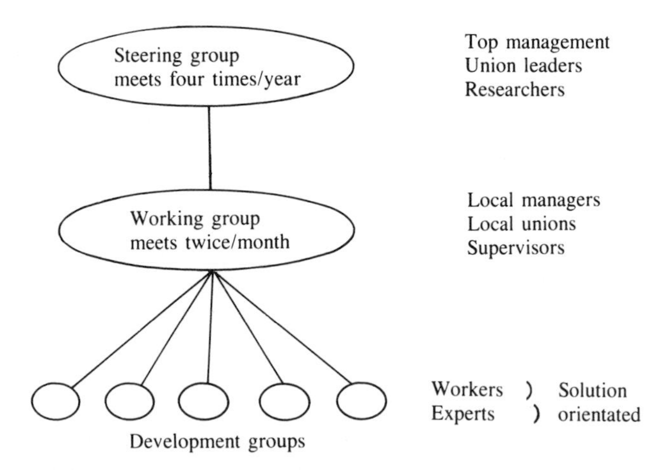

Figure 4.2 The Gothenburg model for solving conflicts

many cases their problem definition has to be changed later. In this process, the researcher often intervenes and provides his 'scientific' definition.

However, there are three possible errors:

1. The problem definition is wrong due to a lack of analytical methods (JIT, lean production, etc).
2. The problem definition only takes into consideration the view of senior management and the researcher.
3. The problem definition sees the technology as fixed and not possible to change. Therefore, the problem is in the social system.

In general, design-oriented research has been conducted with a strategy related to problem definition in the social system. The strategy has often been a step-by-step approach to push specific actions through the organization. Often, the acceptance of this has been low among large groups of middle managers and workers. Lack of understanding of the two concepts, compliance and incentives, has made the research less efficient.

To avoid such problems, the participation process has to be efficient without too hard a control of the process. There is also a need for a method to handle conflicts, which always occur in design-oriented research. These conflicts are frequently mentioned in reports but their analysis and strategies for solving them are seldom documented. Often participation will be seen as the solution and nothing else. Practical experience shows there is a constant struggle during design-oriented research and many negotiations take place. The 'Gothenburg Model' can be used to solve conflicts.

The Gothenburg model is based upon the fact that the steering group has agreed on the types of goal for the organization:

- social goals (meaningful work tasks, autonomy, incentives etc);
- technical goals (capacity, flexibility, utilization etc);
- economic goals (cost, income, transfer price setting, productivity measurement etc).

The design of a production system has to sustain technical, social and economic goals. In practice, it is impossible to reach a design that totally realizes all goals, and a compromise has to be accepted.

In several projects with which the author (Karlsson) has been involved, it was found that either there has been a lack of goals or, when there are goals, they have not been operationalized to a point where they are measurable. Lack of technical goals is very common. For example: capacity, flexibility, reliability, quality and production planning all have to be described in measurable goals. Social goals such as technical autonomy, administrative autonomy and variation within the work task must also be described in terms of measurable goals. Economic goals such as cost and income must be described in different areas, and concepts such as transfer price setting and activity based cost accounting must be involved together with cost accounting. The cost of qualifications is a factor because some designs necessitate extremely high costs for education of the personnel.

In the Gothenburg model, the different goals are given to the working group to reformulate them into more concrete goals with which a development group may work to satisfy different solutions. These solutions are reached in close cooperation between workers and experts to achieve as much participation as possible. Gustavsen's Gothenburg model is a dialogue model on lower levels and a more traditional sociotechnical model on higher levels. It has been proved to resolve conflicts and if a conflict cannot be resolved in the development group, it goes on further to the steering group. This provides good control because certain conflicts cannot stop the process for longer than 3 months.

When we have used a reference group instead of a steering group, the projects have collapsed because there have been no conflict handling mechanisms. There are other models to solve conflicts (Varela, 1971) but, in the author's opinion, they are more time-consuming.

Case studies and action research

Case studies have been largely the basis for design-oriented research and 'star cases' have been described in several reports. However, researchers describing such cases have a tendency to see the system as 'closed' in that the best solution is not an 'open system' solution. When working with open systems theory the perspective must be broadened to take into consideration changes in the environment. The generality of the 'best solution' is limited to a relatively stable environment where the flexibility demands can be measured, whereas in more unstable environments the 'best solution' is of little value. In industrial systems the environment is, in reality, not as unstable as for other organizations. It is also obvious that organizations designed to cope with an unstable environment will not have the generality dilemma.

The main problem with the case study is that the researchers are focused upon success in their documentation and not upon the reasons why certain failures occur. The case study approach needs to be described more in relation to:

- environmental demands;
- detailed analysis of the organization before the change;
- detailed analysis of social, technical and economic goals;
- detailed analysis of why certain actions will be taken;
- detailed analysis of success — failure in relation to actions and goals.

A great deal of criticism against case studies has been expressed due to the lack of participation and democracy, but Swedish experience will not support this criticism. In

Sweden, it is more the lack of analysis in the area of leadership and incentive that can be criticized.

The action-research approach needs a specific method of analysis to distinguish between the actions of the company, the employees and the action researcher. In the traditional Scandinavian action research, the researcher has often been very active and more or less has pushed through his particular idea, selling it to supervisors, union leaders, workers, top management, quality personnel, maintenance personnel, accountants and materials administration personnel. This persuasion process, that is vital in action research, is seldom mentioned in action-research reports.

Many of the techniques used in action research have been chosen by the researcher to suit specific situations in the process. It is often by observations that the researcher understands the problems he has to solve to get the process running. In his mind he therefore classifies his unstructured observations, so that he can choose a 'suitable action'. What to observe, how to classify observations, how to gather them together to form a problem, how to match problems with action is seldom mentioned in action-research reports. The author's experience of action research has alerted him to the need to observe, catch and solve problems early by presenting 'scientifically justified' actions.

The Swedish sociotechnical approach in practice

Much of the literature covering the Swedish approach has been about assembly, and in the 1970s jobs that were classified as boring and unskilled were often associated with assembly. The Swedish National Board for Technical Development then started two programmes, one to create an alternative to the assembly line and the other to replace assembly jobs by automation and robot technology. As project leader of the project 'Alternative production systems to line production' between 1978 and 1982, the author worked very closely with the automation research group and exchanged results during that period.

During the project it was found to be very difficult to utilize flexible automation and very often it incurred too high a cost in the equipment when it was technically possible to automate assembly operations. Instead, automation was economically more effective at earlier stages of the production process such as forging, casting, stamping, welding, milling, turning, drilling, grinding, painting etc.

During the period 1978–1990, the total number of workers in Swedish engineering industry decreased by 20 per cent but the proportion of assembly jobs changed from 40 per cent in 1978 to 50 per cent in 1990. This was caused by a combination of increased automation of non-assembly jobs and greater complexity of products to be assembled. At the same time there has been an increase of jobs with very short work cycles that have been difficult to automate, such as feeding machines, etc. Such jobs often create ergonomic problems. In the literature, there are few examples of sociotechnical approaches to the design of jobs with short work cycles but today these jobs are reorganized in Sweden with the goal of creating more interest, mainly by job enlargement, job enrichment and teamwork. By this reorganization of short cycle jobs there has been a decrease in the ergonomic problems as the jobs have been made more flexible. An early example of this approach is the chemical industry where there has been a long tradition of using autonomous work groups since the beginning of the 1970s. These groups are still in practice as a standard work organization in the industry. In some types of the engineering industry there has been a high level of automation during the end of the

1980s. These industries have adapted autonomous work groups as an organization to integrate both production and maintenance in the same group. Often these groups are working with the Japanese TPM (total productive maintenance) concept.

Practical knowledge about systems design is held by consultants but is rarely documented. On the other hand, action researchers, and other researchers working with practical design of organizations, have not documented *how* they work, rather they have described the effects and focused on participation. The form of publication of such work is through 'star cases' which describe 'the best solution'. All design-oriented research, with the exception of the dialogue model, can broadly be classified as a 'star case strategy'. There are two main disadvantages of such a strategy: first, the generalizability of star cases and, second, the time consumed. Much design-oriented research has been conducted from the perspective of practical problems (concerning new technology, new control systems etc). However, inducements for people to accept the participative social organization associated with such technical systems, such as autonomous work groups and more meaningful work tasks, are often neglected.

Productivity

There is a lack of productivity analyses reported from the Swedish sociotechnical model; this is because the companies wanted to keep this data confidential. However, most of the Swedish experiments have resulted in a number of improvements such as:

1. Labour hour reductions of often 20 per cent or more.
2. Decreased leadtime (work in progress).
3. Increased quality.
4. Decreased absenteeism.
5. Decreased labour turnover.
6. Increased delivery security.
7. Decreased cost for distribution due to decreased leadtime.
8. Increased flexibility in variants and volume.
9. Increased flexibility in working hours.

The experiments have also resulted in higher costs such as:

1. Higher salaries due to increased competence.
2. Increased cost of training.
3. Greater investment in parallel equipment.

The closure of the Volvo Uddevalla car factory, one of the showcases for the Swedish model, has been widely interpreted as a serious drawback to the sociotechnical approach. In fact the Uddevalla plant was closed because it was more profitable to close the company's smallest plant when sales declined. However, the Uddevalla concept has proved to be successful when the cars are designed so that most of the subassembly is carried out at the suppliers and the relationship between subassembly and final assembly is such that the final assembly is 25 per cent of the total assembly and that variant flexibility is needed.

Proof of the Uddevalla concept's suitability is its use to build military battle vehicles at Hägglunds within the ASEA Brown Boveri group. These products have a lot of variants and an assembly time twice as long as for a passenger car, but prove that the sociotechnical approach is still being applied by industrial engineers in Sweden and in areas other than car assembly.

Identified disadvantages

The Swedish approach to match new technology and organization has mainly focused on job enlargement, team work and a democratic leadership style. The disadvantages of this approach can be summarized according to the following five categories:

1. When the company has unclear productivity goals there have been problems to achieve productivity.
2. When the incentive structure has been weak, especially as Sweden has very little difference in wages between unskilled and skilled workers.
3. Where individual competence has not increased due to the job enlargement approach and the organization has not grown in competence, in which case a job enrichment approach would have a much higher effect on competence growth.
4. Where specific competence at the local work place has been favoured instead of general competence. (This approach is opposite to the Japanese approach where both general and specific competence are favoured.)
5. Where the job enlargement approach has lead to problems in the long run when trying to adapt to the flexibility demands outside and inside the company.

In summary, Swedish companies have to increase job enrichment and general competence among their personnel or they will suffer from inflexibility and lack of productivity. Salary and other incentive systems have to be better structured to suit increased competence.

Recent developments

During the late 1980s and early 1990s the sociotechnical approach was extended in various ways.

Technology analysis for R&D organizations

The focus in design-oriented research has been on technical systems such as machines in a production organization. When it comes to R&D organizations, however, the focus has been on computer-aided design (CAD) equipment in relation to work roles.

As the product itself has different technologies, it follows that the work operations in R&D organizations are influenced by the technology used in the product. The author has tried to use this approach to design engineering work organization where operations in R&D organizations are often determined by the technical barriers that need to be overcome.

Using this approach seems to be appropriate and the comparison of social, economic and technical goals with technical barriers is very interesting. Many technical barriers are caused by lack of concrete goals so inclusion of product technology in the technological analysis seems to be necessary in R&D organizations.

Engineering work design

Towards the end of the 1980s and at the beginning of the 1990s, more focus has been directed increasingly on the reorganization of engineering work, especially for product development. Within the car industry, competition from Japan has forced both Volvo and SAAB to decrease their leadtime for new products. At the same time, quality demands from their customers have increased. Almost all industries in Sweden have similar problems.

The sociotechnical approach has been used to create more efficient engineering organizations. Team work, project organization and task groups have been used in combination with more structural methods such as concurrent engineering. The theories developed by Pasmore (1988) and Pava (1983) have been put into practice. Studies by Trygg (1991), Lundqvist (1994) and Liander (1994) have shown that engineering work, and especially product development, represent areas where it is important to match the social system with the technical one in order to achieve effectiveness. Problems that have to be solved can be characterized as deliberations (Pava 1983) and these deliberations (often very specific problems that are possible to discover at the concept level or at an early stage of the project) are solved by a forum of different engineers with a specific knowledge. Most technical problems are solved in the social system according to Liander (1994).

Accounting practice

In all design-oriented research there has been a tendency to prove the productivity in the new design and often behavioural scientists have tried to prove that work satisfaction will improve productivity. However, there does not appear to be any evidence to support this. Productivity can be divided up into inner productivity (doing things right) and outer productivity (doing the right things). Work productivity is nowadays of less value as the cost in production often is the capital. Nevertheless, work productivity is important as a factor influencing capital productivity. If, for example, an assembly task takes 50 per cent less time to accomplish, the flow increases by 100 per cent, and the capital cost decreases by approximately 50 per cent. These are important relationships. Modern production that is market oriented can be more profitable by adopting higher pricing for prompt delivery of a custom-made product. The higher price for flexibility can still give a high productivity even if the cost of production increases. Older accounting practices were not taking this into consideration but newer activity-based cost accounting as described by Kaplan (1986) has made it easier to prove productivity in design-oriented research. Still, there are often problems to redesign economic control systems.

During recent years the focus for increasing productivity has transferred from direct labour hours to capital costs (inventories and equipment). The product design that stands for 70 – 80 per cent of the production cost has also been focused upon and most Swedish companies nowadays analyse new products in relation to manufacturability using techniques such as design for assembly (DFA) and design for manufacturing (DFM) (see Lundqvist 1994). In the 1990s, 'activity based costing' (ABC) has been adopted by most of Sweden's multinational companies and some of them have also gone over to 'activity based management' (ABM).

Due to the market demand for increased flexibility within production, some companies have started to adopt transfer pricing between sales and production. In reality, this means that the production cost can increase if the sales price or the market share increases. This has been necessary as the cost of extra machines and personnel that the flexible manufacturing systems normally need has to increase along with total productivity (Lindér, 1990).

To compare one production system with another zero-calculation, developed by Karlsson (1990), has been widely used in Swedish companies. To compare different production concepts with zero-calculation (with no losses in people and capital) for a fixed product has been very fruitful when designing sociotechnical systems (Karlsson, 1993). Zerocalculating is one of the few tools available to create a lean production. It is therefore very relevant to the following discussion on Japanese management.

Japanese and Swedish management

The two dominating perspectives applied in design-oriented research are the sociotechnical approach and the organizational development approach. Other perspectives also exist such as the 'democratic dialogue model' (Gustavsen *et al.*, 1991) and the Japanese 'kaizen' perspective (Imai, 1986). The democratic dialogue is a sociotechnical approach which focuses upon the participation process. It is not oriented towards a specific problem or goal, rather it is more a way of defining problems. The dialogue model has been used in a fairly small number of workplaces in industry compared with the traditional sociotechnical approach. The kaizen perspective has been applied in Swedish industry during the last few years. Its focus on 'small step problem-solving' is complementary to the traditional sociotechnical approach. It is a perspective rooted in bureaucratic theory, where legality as acceptance of the goal is the important factor. Despite being widely used in practice it is seldom described in the literature, although cases are described where different types of bureaucratic theory are used based on compliance and legality (Tichy and Devanna, 1986).

Japanese management ideas, such as JIT, total productive maintenance (TPM), lean production, total quality management (TQM) and kaizen, have been widely used in Swedish companies together with the traditional sociotechnical approach. They fit in well with Scandinavian management with its emphasis on teamwork, participation and responsibility.

Some people in Sweden emphasize the distinction between Japanese lean production and the Swedish sociotechnical approach (Berggren, 1993). After working as an action researcher and consultant in Swedish industry for over 20 years on production systems design and the organization of R&D work, the author is, however, of the opinion that the Swedish sociotechnical approach in the form of the Gothenburg model (not the democratic dialogue) can easily be combined with the lean production concepts of TPM, JIT and kaizen. On the other hand, Swedish management style cannot be replaced by Japanese management and many Japanese observers are convinced that the Swedish management style only needs to adopt some parts of Japanese management.

Many companies have started up programmes to change the company so that both autonomous work groups and lean production can exist together. One example is ASEA Brown Boveri with their T50-program (Berger, 1993). Here, Japanese management has highlighted more goals to achieve and complements Swedish management where goals are often unclear to both production personnel and supervisors.

Future outlooks

A new approach to the sociotechnical tradition in Sweden is growing in strength. The old tradition from the 1970s concentrated on participation law, worker satisfaction, decreased absenteeism and labour turnover. Many companies in the 1980s have had an unfavourable experience of the sociotechnical approach due to the fact that they never paid attention to the incentive system nor established clear productivity goals for the groups. However, in the long run, there must be a change from job enlargement to job enrichment through a change in competence. Sweden has relatively few skilled workers due to its industrial structure and so a great deal of training has to take place within the companies. There will also be a change of focus from assembly jobs to those in highly automated industries and industries with short work-cycle jobs. It is important to be able to design production organizations for this type of industry, but it is more a question of the design of production concepts than of organizations.

White-collar work, especially that carried out by engineers in product development, is an area where more attention needs to be paid regarding the technological analysis of the work. The concepts and theories are weak in this area although Liander (1994) has developed some of the theories of Pasmore (1988) and Pava (1983). The different unions for white- and blue-collar personnel work towards competence growth as blue-collar jobs have been reclassified as white-collar ones after organizational changes. The dilemma is that very few young people want to educate themselves to be skilled blue-collar industrial workers. One reason for this has been that there is little difference in salary between skilled and unskilled workers. In the future, therefore, Sweden really needs one union for white- and blue-collar personnel and a bigger difference in salary so it will be profitable for workers who educate themselves.

Productivity and management accounting need to be linked more tightly to the sociotechnical approach. Developing total productivity measurement methods is important and some basic models have been developed by Karlsson (1990) such as the productivity circle. However, there is still a lot more to do before a practical model is created. To develop the ABC and ABM concepts for budgetary and economic control is not easily achieved, but it is necessary to be able to calculate different sociotechnical concepts for both white- and blue-collar workers (Rosander, 1994; Börjesson, 1993).

An area that is extremely weak in the sociotechnical approach is the development of methods to control the change process in relation to a new production system or engineering organization (Naschold, 1992). The sociotechnical approach must be broader than the original behavioural theory. Technology analysis, change process control, economic calculation and productivity are areas that need to be more developed into practical methods. Other areas that need to be developed are incentive and competence structures.

In the 1990s, since senior management and unions have achieved a better understanding of Japanese production, they tend to mix Japanese lean production and the Swedish sociotechnical approach so that there is a strong tendency to group work organization and lean production techniques. In summary, therefore, the Swedish sociotechnical approach, combined with Japanese management and broadened to 'concepts', is a possible way to design jobs in relation to technology in the future.

References

Berger, A. (1993) 'Organising manufacturing development. Traditions, assumptions and managerial implications — A status report', licentiate dissertation, Gothenburg: Chalmers University of Technology.

Berggren, C. (1993) Lean Production — The End of History?, *Work, Employment and Society*, **7** (2), 163–188.

Börjesson, S. (1993) Activity based approaches — Is activity information useful for resource control and performance improvements?, in: Karlsson, C. and Voss, C. (Eds), *Management and New Production Systems; The 4th International Production Management Conference*, London: London Business School/EIASM, pp. 121–138.

Engström, T. (1983) 'Materialflödessystem och serieproduktion, doctoral dissertation, Gothenburg: Chalmers University of Technology.

Gustavsen, B., Hart, H. and Hofmaier, B. (1991) From linear to interactive logics: Characteristics of workplace development as illustrated by projects in large mail centers, *Human Relations*, **44** (4), 309–322.

Imai, M. (1986) *KAIZEN — The key to Japan's competitive success*, New York: McGraw-Hill.

Kaplan, R.S. (1986) *Accounting Lag: The Obsolescence of Cost Accounting Systems. The Uneasy Alliance — Managing the Productivity/technology Dilemma*, Boston: HBS Press.

Karlsson, U. (1979) *Alternativa produktionssystem till lineproduktion*, Gothenburg: Chalmers University of Technology.

Karlsson, U. (1990) *Production Measurement and Improvement*, Gothenburg: Chalmers University of Technology.

Karlsson, U. (1993) 'Zero Calculation and Lean Production', paper presented at the Eighth World Productivity Congress, Stockholm, 23–27 May.

Liander, K. (1994) 'Sociotechnical analysis of product development work', doctoral dissertation, Gothenburg: Chalmers University of Technology.

Lindér, J. (1990) 'Värdering av flexibel produktionsorganisation utifrån sociotekniska principer', doctoral dissertation, Gothenburg: Chalmers University of Technology.

Lundqvist, M. (1994) 'Process management in product development', licentiate thesis, Gothenburg: Chalmers University of Technology.

Naschold, F. (1992) *The LOM Program*, Stockholm: Swedish Work Environment Foundation.

Pasmore, W.A. (1988) *Designing Effective Organizations — The Sociotechnical Systems Perspective*, New York: Wiley.

Pava, C.H.P. (1983) *Managing New Office Technology; An Organizational Strategy*, New York/ London: Free Press/Collier Macmillan.

Rosander, K. (1994) 'Design of production systems for batch production in small volumes', doctoral dissertation, Gothenburg: Chalmers University of Technology.

Tichy, N.M. and Devanna, M.A. (1986) *The Transformational Leader*, New York: Wiley.

Trygg, L. (1991) 'Engineering design — Some aspects of product development efficiency', doctoral dissertation, Gothenburg: Chalmers University of Technology.

Varela, J.A. (1971) *Psychological Solutions to Social Problems; An Introduction to Social Technology*, New York: Academic Press.

van der Zwaan, A. (1975) The sociotechnical systems approach, *International Journal of Production Research*, **13** (2), 149–163.

5

'Technikgestaltung' (Shaping of Technology) and Direct Participation; German Experiences in Managing Technological Change

Erich Latniak

Introduction

In order to introduce the arguments which will be developed in this chapter, it is useful to give some information on the perspective of work from which they are derived. First, it should be mentioned that this chapter will focus only on the German discussion of *Technikgestaltung* ('shaping of technology'; 'shaping' is used in the following as the German word *gestalten* covers the whole chain beginning with design of technologies, developing, implementing, adapting and using them at the workplace), and on the development of symbiotic approaches in Germany. The problems of 'Technikgestaltung' first evolved in the early 1980s as a side effect of the diffusion of new information and communication technologies (ICTs) that was beginning to take place. Symbiotic approaches, which have a longer tradition in Germany, were seen as possible solutions to these problems. At that time, they had been developed further as a critique and in opposition to 'technocentric' concepts of computer integrated manufacturing (see Brödner, 1985: 61–116). What seems to be specific for Germany is that the international discussions on comparable problems have been partly adopted, but this had little influence in practice. This might be due to the fact that the manner in which the adaption and diffusion of new ICTs has been managed is widely determined by social aspects and regulations which are specific for an individual firm or types of firms on the one hand, and which are determined by legal regulations and industrial relations on the other. Another reason is the ongoing dominance of technically oriented solutions for problems of productivity and control, which despite all scientific discussions is widespread still.

A second point is that the questions and problems presented in the following sections derive from a specific position integrating scientific research, development, and consulting in an action–research perspective. The main problem of this perspective is to find a path that realizes both objectives in one process: to support enterprises to reach higher productivity, flexibility and product quality by 'bringing back work to the factory' (Brödner and Pekruhl, 1991) and to improve working conditions and broaden the scope of the workers' tasks at the same time. Evidently, there is no lack of knowledge, ideas, concepts and methods for different purposes, but there is a lack of knowledge on how to increase the diffusion of these concepts and approaches and how to support enterprises in applying symbiotic approaches effectively.

A third point to be preliminarily mentioned is that the chapter cannot present a coherent theory or integrational approach. Its main objective is to present some elements which might be fruitfully developed further on and to ask questions that necessarily have to be answered if a further diffusion of symbiotic approaches is intended.

The following section will first discuss aspects that can define symbiotic approaches. It will then give information on the situation in Germany and show the methods selected to be used in symbiotic approaches. This will be illustrated by some empirical findings on implementation processes, and focusing on their difficulties.

Symbiotic approaches

International comparative research on successful methods of production during the late 1980s demonstrated that there are production systems reaching much higher productivity, reduced lead time, better product quality standards, and increased innovation capacity than achieved by most 'traditional' production systems in Europe (Womack *et al.*, 1990; Dertouzos *et al.*, 1989). Investigation into the reasons for this fact led to a strategic shift of perspective: technical aspects are still one, but not the only determining reason for success. At the time of writing, it is widely consented that the technology in use must sufficiently support human work rather than replace it. What is necessary is an integrational or 'symbiotic' approach: to reshape organization, qualification, and technology as a coherent whole. The term 'symbiotic approach' will be used, therefore, to characterize approaches of implementing technologies by integrating technical, organizational, and individual aspects in order to develop a successful production system. Technology, organization, and qualification are to be developed in an integrated manner.

There are several problems to be solved for a fruitful integration. Scientists are used to thinking about the process of shaping according to different professions and learned methods. It is not certain if there can be only one coherent theoretical basis for symbiotic approaches and all the methods to be used. The present impression is that there are manifold of methods and tools to be used within symbiotic approaches but they evolve from different sciences and methodologies. So, a symbiotic approach in practice will be a modular approach. It is an aspect for further discussions if an integrated theoretical basis can be developed at all and which axioms could be the basis of this theory.

A second aspect is the broad scope of practical problems for which an integrated approach could be a solution. This approach might have to cover solutions for the shaping of a single work task, the planning of the organizational design of an integrated semi-autonomous work group, the technical restructuring of a production line, or the design of a machine. In the author's opinion, a symbiotic approach can only be a modular one. A modular concept could be adapted best to the specific needs of individual settings in the firms. To reach an integrated perspective, each of the single modules should correspond to some basic principles. Some of these principles discussed in Germany will be presented in the following sections, for example:

- According to experiences with restructuring processes, it might be the most effective way to start restructuring by solving the problems of the organization of production on the shop-floor. The restructuring is primarily oriented to work tasks.
- This starting point directly leads to the idea that machinery is to be understood as a tool to support work as efficiently as possible.

- The work organization should be structured in an 'object-oriented' fashion, i.e. families of products should be produced integratedly by semi-autonomous groups of workers who are planning the details of production.
- The whole process of restructuring must be planned and conducted in a participational way.

A last problem to be mentioned is a consequence of limited knowledge. All concepts and approaches are closely related to the empirical areas where projects and concepts are developed. Therefore, the academic discussion in Germany is driven by efforts to support the production based on skilled work and by mobilizing the competences of the workforce. The 'blueprint' for most of the methods and concepts developed is the German machine tool industry which has a high degree of skilled workers and relatively complex production structures. This may cause a bias in perspectives.

Interdisciplinarity

At present, there is no single profession and no single science which can manage all the problems of organization, technology, and qualification sufficiently. The specific competences of engineers, social scientists, etc., are still needed to find and apply adequate solutions in practice. Evidently, it is necessary that symbiotic approaches are interdisciplinary approaches. Therefore, the realization of symbiotic approaches in practice leads to a team-based project structure where all these necessary competences should be represented.

The present impression in Germany is that many of the symbiotic approaches are not 'mainstream' approaches of the sciences involved. Many of the concepts developed stay 'in between' professional borders, e.g between psychology, sociology, and engineering. The research groups developing symbiotic approaches seem to play the role of 'pivot players' (as discussed in game theory) transferring knowledge into different sciences.

Publications, such as that by Brödner (1985), have resulted from close cooperation with social scientists at the Berlin Science Center, but the results were less adapted to engineering, due to the dominating CIM-concepts at that time, than to sociology. For work psychologists it is difficult to be represented in conferences of psychology: the scientific main stream has a different orientation in testing methods and scientific recommendations. Therefore, closer cooperation with the engineering sciences evolved.

If the impression is correct, this fact leads to a recommendation for symbiotic approaches: the solutions to be developed have to lead to more visible practical effects in use than achieved by traditional approaches. As these approaches stem from different sciences, they will be accepted in academia only if they fit the recommendations of the different sciences involved.

Changing role of participation

According to the theory that technology, organization and individual aspects have to be taken into account in an integrated manner, forms of direct participation in development and implementation of technology, supported by vocational training measures, are crucial elements of symbiotic approaches.

It is obvious that the evaluation of the role of the workforce in production is about to change. Although in Sweden during the 1970s, problems such as high degrees of absenteeism (while labour markets could not provide skilled employees) resulted in the development of new approaches in the organization of production, there is a lot of evidence for the fact that the increases in productivity to be gained today derive from the manner in which production is structured and how workers interact in the production (Pieper and Strötgen, 1990). If the workers' and employees' roles in production has to change for economic reasons and if flexibility, quality, productivity, and higher innovation rates are not predominantly influenced by technology but in fact stem from the efficient use of the workforce, it is evident that the employees involved in change processes have to undertake an active role.

This has several and far-reaching effects on the implementation process itself. An implementation by using a top-down-strategy based on management decisions is not an adequate solution for purposes of participation. On the other hand, if the implementation process is based on participation, the objectives defined by management will be changed, modified or broken into pieces and selectively realized. The degree of participation and decision is to be cleared, the process of implementation will cause much more debate and turmoil than before.

Keeping this in mind, a broader use of participational means indicates the belief that implementation and process aspects have to be taken into account in order to make the approach more effective in practice. The limited diffusion of symbiotic approaches might be a result of ignoring the preconditions, limits and obstacles of change processes inside the firms. Focusing process aspects and the organization of the change process itself as elements of a symbiotic approach will provide a more dynamic view on these approaches.

Internal restrictions and obstacles

The main purpose of symbiotic approaches, seen from a management perspective, is to reach a higher degree of rationalization and efficiency in production. At least, two main restrictions of these efforts can be named in advance. The first obstacle is a restriction of time. While the changing of a machine can be managed relatively quickly and in a prescribed way, changes in organization, habits, action, and values of people working together and their interaction can only be changed quite slowly and over a long time. The second obstacle is that every change in machinery or technology will cause a change in the social structure of production. The power and influence of specific divisions being involved in the implementation of a technical system for production planning may change drastically. Therefore, the individual interests and resources of management and employees in these divisions are touched and they will begin to support or oppose the project according to the chances of success and individual influence they may gain. The concept of 'Mikropolitik' has been developed to describe these internal restrictions for successful processes (see the section on Mikropolitik below).

Diffusion and practical use

Empirical findings carried out during the early 1990s reveal that the diffusion of symbiotic approaches into practical use is still very limited. Recent research of the IAT provided, for example, a low degree of diffusion of team-based production structures (i.e. integrated work tasks in production and in planning and a high degree of responsibility of the workers for a specific task) (Kleinschmidt and Pekruhl, 1994: 29). Based on a representative questionnaire survey on the diffusion of group-oriented work structures in Germany, only 6.9 per cent of the employees in industry are actually working in structures of this kind. There was no evidence for the belief that in smaller enterprises group-oriented structures would be used broadly, but not classified as group work; the greater the number of group structures, the larger the enterprise. Another aspect must be highlighted: while most of the group-structures have been initiated 3 or more years ago (81 per cent approximately), only 11 per cent were initiated 1–3 years ago and only 9 per cent during the last year (Kleinschmidt and Pekruhl, 1994: 39).

A study on the diffusion of new production concepts and market strategies in industry in North Rhine–Westphalia in 1990 had analysed that there is no direct nexus, but rather a contingent relationship between market strategies of firms and the organization of their production (Hennig and Pekruhl, 1991). Flexible specialization (production of small batches according to the needs of the customer) as a market strategy does not lead to team-based production structures. It could be demonstrated that the dominating strategy is a reactive and incremental type of implementation, e.g. of new ICTs in the industries, and lacks organizational concepts and perspectives.

These research findings indicate in particular that symbiotic approaches are well documented in academia, but have less influence in practice. In order to increase the diffusion of these methods and approaches, a further investigation into the implementation processes and conditions of failure seems necessary.

Developing symbiotic approaches in Germany

Until the mid-1980s there were few documented projects in which a company tried to use symbiotic approaches with its own resources. Even now, the diffusion of these approaches is quite low, while the discussion on 'lean production' became popular and group-oriented work structures are broadly discussed. At least, most of the large automobile plants and the biggest automotive producers in Germany have been trying to implement group structures in production for several years now (see Roth and Kohl, 1988; Binkelmann *et al.*, 1993). But two points should be noticed:

- The group concepts and the tasks to be fulfilled by the groups vary significantly even within the same factory. Only a few semi-autonomous forms of group work are among these concepts. The stability of these organizational approaches has not been investigated yet.
- The underestimation of non-subsidized projects of firms is due to the fact, that these projects often lack resources for external demonstration, etc.

The best way to describe the changes in discussion on symbiotic approaches in Germany is to use publicly funded research programs and their results as examples and guide. In the following sections, we will give some background information and some examples of methods developed in projects to illustrate changes and 'state of the art'.

Traditionally, in Germany, the REFA (Verband für Arbeitsstudien- und Betriebsorganisation; German Association for Ergonomics and Work Studies) system has been the dominant method of analysing and shaping production processes, rationalization, and improving productivity. The method has been developed and adapted by engineers over many years since the early1920s. In the beginning, it focused on a single work place and tried to avoid unnecessary moves and to reduce the resources necessary for specific tasks. Using REFA was, at least in unionist discussions, a synonym for worsening working conditions, rationalization and losing of jobs on the one hand, and for an increase of productivity and scientific management on the other hand. The REFA system describes six necessary, discrete, and sequential steps of a shaping process to be followed in *each* shaping process in order to reach the aims defined. These are:

1. analysing the situation in the beginning;
2. defining objectives and tasks;
3. making a concept of the work system;
4. detailing the concept;
5. implementing the concept;
6. using the system.

These steps were defined in more detail for different purposes (complex production systems, organization of the factory, organizing bureaus, organizing production) (REFA, 1987; REFA, 1991a,b,c). This static process model was, and still is, the German state of the art in engineering and shaping of production system.

The early international experiences (e.g. Tavistock, Scandinavian approaches) were not immediately adapted in Germany. This situation changed at the beginning of the 1970s. Even though this was not the central aspect of the programme, the federal research programme 'Humanisierung des Arbeitslebens' (HdA) can be seen as a changing point for the discussion and an initiation for a working up of international experiences on organization and work structuring.

'Humanisierung des Arbeitslebens' (HdA)/'Arbeit und Technik'(AuT)

HdA stressed aspects of working conditions and the security of working life when it started in 1974. It soon became a well-known and well-subsided federal investigation-programme. At present, more than 1000 projects have been financed within these programmes dealing with working conditions, health problems in working life, and work structuring.

In the beginning, work structuring was a problem initiated by automatization projects within several large firms; it was not at all a common topic to German industry. Rethinking the organization of production was indicated by interface and buffer problems of highly automatized production lines. The limits of these tayloristic structures now became evident after a long period of success. The efforts to substitute human work by the use of machines as far as possible led to an inflexible and highly vulnerable production system. Therefore, in some industries, especially in the automobile industry, first approaches of restructuring and regaining flexibility were initiated. As a consequence of this limited consciousness of automatization problems, important parts of the German industries did not take an active part in the HdA programme at that time.

One of the main levels of HdA's activities was the shop-floor level at which the projects were done; the other level was the intermediary level of bi- or tripartite (including policy makers) discussions among unions and employers' organizations. It should be noticed that a cooperative way of managing or even discussing questions of industrial restructuring in Germany has been a strategy of minorities within the organizations of unions and employers. So it is not astonishing that the strategy of most of the workers' councils until the early 1980s was primarily oriented to avoid job losses as a consequence of rationalization or automatization processes and to concentrate on wage problems. There was no cooperative culture of industrial relations at the shop-floor level. The orientation within the largest unions began to change in the early 1980s, when the first conceptual discussions and several initiatives were started. Therefore, thinking about an integration of increased productivity and increased working conditions was far beyond the horizon of the majority positions of the interest groups involved. HdA and AuT, especially their consulting groups, were steps towards a cooperation of at least parts of the unions and parts of the employers' associations (see HdA-Projektträger, 1981; FFES *et al.*, 1982; Hanewinkel, 1986; Volkholz, 1991; Peter, 1990)

Several industries took part in the programme and, though they were a minority of projects, a couple of well-documented success stories of restructuring processes and the application of symbiotic approaches in the 1980s should be highlighted:

Hoesch Kaltwalzwerk, Dortmund

In order to manage problems with high fluctuation rates of employees and absenteeism (which could not be regulated due to an inflexible production system) a concept of group-oriented structuring was implemented in two steel mills fostering homogeneous tasks for the employees. Each group member should be able to use all machines. A close cooperation between groups, workers' council and management is an interesting aspect here (Lichte and Trültzsch, 1986; Lichte and Reppel, 1988; Jürgenhake and Lichte, 1990; Lichte and Jürgenhake, 1992).

Felten & Guilleaume Energietechnik, Nordenham

Beginning in the early 1980s, this project led to a complete restructuring of production. The organization was set to semi-autonomous groups and a production island: 30 islands, with 3–8 workers on each island, producing a variety of product families in smaller batches. The whole process was supported by a training programmme for all the employees involved. The main problem was a high degree of unskilled and semi-skilled workers (Kiehne and Kohl, 1988; Koschnitzke and Köster, 1989).

The HdA programme has been in political turmoil because of its close connections with the unions. This lead to a reorganization of the programme and its political intentions. In 1989, an integrated perspective on innovation covering technology, work organization, and working conditions has been chosen and broadened the scope of activities. This is symbolized by the new name of the programme since that time: 'Arbeit und Technik' (Bundestag, 1988).

To sum up the impressions on the discussions on symbiotic approaches until the early 1980s in Germany, six aspects are evident:

1. Within the HdA/AuT programme, work structuring projects have been in the minority for a long time, and integrated 'symbiotic approaches' have been a minority within these projects. The majority of HdA/AuT projects dealt with health and safety problems.
2. One of the main problems was that the solutions integrating technical and organizational aspects were developed for a single firm-specific problem normally. Even though HdA/AuT can present remarkable success stories of several projects

(FFES *et al.*, 1982), the main problem was the transfer of the project results. The projects lacked an integrated transfer concept.

Meanwhile, it is evident that 'success stories' are insufficient for the dissemination of the results; in some cases, even the know-how transfer within the firms could not be managed sufficiently. In order to solve the diffusion problem, complex project structures involving different firms and research institutes have been launched. But the dissemination of the results concerning organizational problems into everyday practice remained limited.

3. The economic success of the projects has not been pointed out as it would have been necessary. At the end of the financial subsidies for the projects, the efforts broke down in some of the firms, a part of the concepts were cut back to the *status quo ante*. This is due to the fact that for a long time 'Humanisierung des Arbeitslebens' (humanization of working life) has been regarded as a supplementary task and not as one aspect of an integrated approach in modernizing the production structures. Therefore, within the firms' humanization has been a cost aspect and not an investment in productive factors.
4. There was no real integration of engineering and social sciences even within the HdA programme for a long time. Despite some examples, social scientists were not involved in action-research projects, but did their investigations in a complementary research while the main focus was set to the healthiness of working conditions. Technology and work organization remained separate worlds investigated by differently oriented sciences. Only in a few areas (robotics) could cooperation be organized (see Volkholz, 1991: 26).
5. A crucial point of the limited diffusion of team-based work group concepts are wage contracts. At Volkswagen (VW), for example, wage contracts were negotiated separately. In most of the other cases, many problems were induced by the structure of wage contracts: an adequate wage system for group structures in production fostering individual commitment is quite complicated to design.
6. Despite the methods, models, or tools used during the HdA-projects, a coherent approach has not been developed within the programme. Even a comprehensive view of the experiences is missing still.

The most 'durable' results of HdA/AuT were several methods of work analysis, developed by the Berlin research group of Walter Volpert, a work psychologist. VERA ('Verfahren zur Ermittlung von Regulationserfordernissen in der Arbeitstätigkeit'; instrument to identify regulation requirements in industrial work), the best known of these methods, was first published in 1983 and has been further developed since (Ducki *et al.*, 1993; Dunckel *et al.*, 1993; Leitner *et al.*, 1987). It was a complete change of perspectives in work-place analysis because VERA is not primarily interested in focusing the functional aspects of work (as REFA-system does for example) but tries to integrate the perspective of individual work tasks and intends to give advice for an improved organization of work (see Volpert *et al.*, 1983).

VERA and RHIA methods

VERA and RHIA (Verfahren zur Ermittlung von Regulationshindernissen in der Arbeitstätigkeit; an instrument used to identify regulation barriers in industrial work) are psychologically based methods used to analyse work and work tasks. The intention was to develop a theory-based tool for analysing work places, focusing on psychological strain and stress. Time restrictions, information, movement restrictions and overcharge of workers' individual tasks are investigated. The method is based on focused and protocolled interviews and observation techniques. The protocolling standards are scientifically defined and very detailed. For a single work place, a trained investigator needs 1 – 3h to analyse and classify according to the manual.

The theoretical basis of these methods is the 'Handlungsregulationstheorie' (action regulation theory), supposing that:

1. human action is oriented to objectives;
2. human action is directed to external objects to be manipulated;
3. human action is embedded into a social context;
4. human action is a process.

The theory is aiming at developing criteria for 'humanized' or human adequate work and focusing the role of work for individual personality. The methods derived from the theory can be described by following elements:

- VERA/RHIA are focusing on the individual work task as the central element of the shaping process. This is the main difference to the function-oriented REFA system. But VERA/RHIA are integrating technical, movement and psychological factors influencing working action.
- They are designed for measuring and analysing stress and the necessity for cognitive regulation of action. The result is a classification of the work place according to different types of cognitive regulations and strain. The investigator is able directly to derive proposals for improvement (Leitner *et al.*, 1987: 7ff; Volpert *et al.*, 1983; Oesterreich and Volpert, 1986). Five levels of cognitive regulation are defined.
- The main objective is investigating conditions of working action.

Within the framework of action regulation theory, two criteria for human work can be defined. Human work should integrate decisions and cognitive planning processes for each task; therefore tasks integrating cognitive planning processes are positively classified. Disturbances, obstacles and overcharge are to be classified negatively because they are causing strain and stress. The methods have been evaluated (reliability and validity).

The central aspect of these methods is the focus on work tasks as the central unit of investigation into working conditions and work organization. Starting from the analysis of ongoing work processes, improvements in technology and organization can be derived adequately. One main advantage of VERA/RHIA is the fact that these methods start from a psychological theory of work as basis. All recommendations and advice derived from this method are not normative but they are based on a theory and on scientifically proved facts.

'Programm Fertigungstechnik' (manufacturing technology programme)

In 1979, several parts of the humanization programme dealing with robotics and flexible automation were taken into a specific programme with less union influence, and increased — stressing the technological aspects of innovation and rationalization processes. The main aim was to develop machinery adapted to improved working conditions, higher productivity, and in accordance with at least three criteria:

1. diminishing physical exertion;
2. enlarging freedom of action for workers by supporting them with adequate technology;
3. reaching higher skill levels.

Several examples of projects have been published (e.g. Brödner, 1985: 149–161; Dankbaar, 1987: 346ff). For example, the restructuring project at ZF and the concept of production islands.

The restructuring project at Zahnradfabrik Friedrichshafen (ZF)

The main aim was developing a flexible and highly productive automated production system for the manufacturing of several hundred parts in small batches of 50–500 units. The project had a complex cooperation structure involving five research institutes and seven machine tool manufacturers and was aimed at increasing manufacturing without deskilling the workers.

A complex manufacturing system based on principles of group technology and semi-autonomous workgroups could be realized. The experiences in implementing this highly complex system lead to a modified concept with less stressing technical aspects but intensified training and preparation for the workers, which was used in another factory of ZF later.

The concept of production islands

These production islands (Brödner, 1985: 141; Brödner and Pekruhl, 1991: 9–15 and 35–40) organize production according to the following principles:

- A family of parts is to be completely produced within the production island and is the full responsibility of a group of skilled and flexible workers.
- The work is structured to semi-autonomous work groups (embedded in a production environment, but as independent units within the company). A 'congruity' of product and organization is to be realized: one organizational unit, the group, is responsible for the production of one group of products ('object-oriented' structure; the expression is used in a different meaning in software engineering though similarities between both concepts might probably be identified). The group is the producing and the controlling organizational system as well.
- Widespread options for intellectual development have to be realized by an integration of tasks (disposition and planning).
- Production islands are a work-oriented design concept (contrasting with traditional technically oriented approaches such as CIM for example).

The main conceptual gain of these projects is the concept of an 'object-oriented' structure (i.e. product families integratedly produced by a cooperating group of workers) of production organization concentrating responsibility, planning, and production in semi-autonomous work groups to a 'family' of products. This can be seen as a structural guideline in accordance to the task-oriented and theory-based approaches within VERA/RHIA. The development of this organizational concept and VERA/RHIA-methods could be realized in a close cooperation among different projects and research groups funded by the two programmes 'Humanisierung des Arbeitslebens' and 'Fertigungstechnik'.

The production technologies programme has been one of the main initiatives to develop and to foster the diffusion of work-place oriented programming (WOP) techniques. Several projects have been launched to increase the quality of the technology and to spread their application. After a period of 10 years, it can be stated that WOP is a well-adopted approach especially in smaller firms which do not have a software division for the programming purposes. A recent study revealed that more than 50 per cent of all the firms in the German machine tool industry with less than 200 workers are using WOP technology (Ostendorf and Schmid, 1994: 17 ff.). What seems to be a success of a symbiotic approach in the first view must be seen in relation to the missing information on how the WOP technology is used in these firms. An analysis of the organizational structures in which WOP technologies are used is still missing. So, the panel study cited states a decentralization of the structures indicated by these figures, but is avoiding statements on humanization of the working conditions.

Without going too deep into details of each (well-documented) project, four aspects of the HdA and Fertigungstechnik (FT) programme are to be highlighted:

1. Only a limited integration of technical/engineering and organizational/social science aspects has been realized within HdA/AuT and FT, but in some areas of research it has been really fruitful.
2. There is an integrated innovation concept within the AuT since 1989 as its main objective, but still few projects exist that aim at developing an integrated approach. This is a consequence of the changing priorities in financing due to the German reunion. The intention of the Federal Research Ministry to distribute research funding to the new federal states in all programmes to a degree of 30 per cent led to severe cutbacks in funding in the 'old' federal states.
3. There is still no theoretical coherence in all these approaches.
4. There was a tendency to overload single but complex and financially restricted projects by an integration of too many different objectives and project partners.

New aspects: 'SoTech'-programme

Due to political reasons, such as the change of federal government to conservative in 1982 and a change in the perspective of public and scientific discussions ('information and communication technologies' (ICTs) became a new topic), the social-democratic government of North Rhine-Westphalia (NRW), a German federal state, initiated a technology programme fostering participation and investigation of ICTs' effects on working and everyday life. The programme 'Mensch und Technik — Sozialverträgliche Technikgestaltung' ('SoTech') can be seen as a follow-up programme to the early HdA (which continued) but was tailored to the specific interests of North Rhine-Westphalia. About 170 projects so far have been financed.

North Rhine-Westphalia had specific advantages and experiences which facilitated initiating a policy programme like SoTech. The research funded by HdA since 1974 had been carried out in the mining industry and in steel mills to a large extent where a specific and far-reaching form of co-determination of the workers' councils has been implemented ('Montanmitbestimmung'). Most of these mills have been located in North Rhine-Westphalia. This is one of the reasons for the high degree of miners and workers who are unionists and for the relatively strong influence of the unions on the social-democratic government. Another aspect of the concentrated research funding in this federal state is a highly developed research infrastructure dealing with the problems of the humanization of working life as a research tradition. The SoTech-programme could start from this basis but it reformulated the perspective of research according to the changing needs of the restructuring processes of the mid-1980s and the scientific discussions at that time (see Latniak, 1995; Latniak and Simonis, 1994).

Within this programme, the North Rhine-Westphalian government tried to integrate aspects of technology assessment and a strategy of societal discourses on the effects and the use of ICTs. The projects on work organization focused the implementation of new ICTs as critical for working conditions and productivity. The intention was twofold: on the one hand, government tried to avoid time lags in the diffusion and the use of these technologies in order to regain higher productivity within NRW's industry. On the other hand, the intention was to strengthen the interests of people involved in ICTs' implementation by vocational training, information and consulting and by initiating discussion processes. Both objectives were seen as necessary for the success of projects and programme (see von Alemann and Schatz, 1986; von Alemann *et al.*, 1992; Latniak and Simonis, 1994).

The new aspects of the SoTech projects on work structuring and ICTs' implementation were at least fourfold.

First, the main focus of the projects was on 'softly' influencing the implementation by mobilization of the people involved and by teaching, training, and initiating learning processes. Hence, questions of vocational training, information and consulting in change processes were fostered. Direct participation became a central means in implementation processes additional to delegative representation of workers' interests via workers' councils. On the other hand, it became evident that the participational process cannot be structured in detail and that the results of implementation processes are based on compromise and bargaining within the firms.

Second, focusing the implementation processes led to a shift in perspectives: the more detailed the investigation of these processes became, the more evident became the fact that in many cases the implementation ICTs were used to break up petrified organizational structures. ICTs in this view became a means to initiate a reorganization of the production processes. The perspective of investigation became process oriented (Manz, 1990; Ortmann *et al.*, 1990; Rolf *et al.*, 1990; Behr *et al.*, 1991; Christmann and Schmidt-Dilcher, 1991).

Third, a further result was that in many cases, especially small- and medium-sized firms, organizations were not able to solve their problems without external support. The enterprises often had a lack of problem consciousness and if they realized their difficulties, they were not able to react accordingly due to a lack of time (quick changes), a lack of financial or human resources, or a lack of know-how in managing social processes inside the firm. This kind of problem led to a complex model of projects trying to integrate several firms, researchers or consultants, and vocational training centres in order to solve common problems.

Fourth, together with the idea of societal discourses, the 'passing over' of the firm's boundaries led to a more complex view of 'shaping' work organization and technology including aspects of technical standards (ISO or DIN-standards), vocational training aspects, and to a different view of diffusion, thus taking external restrictions of the shaping process into account.

As an example for a specific approaches developed within SoTech, STEPS, a method of participational design and development of software, is presented in the following section (see Brödner *et al.*, 1991: 59–94).

STEPS (software technology for evolutionary participative system development)

Based on an investigation of the Scandinavian approaches in software engineering (Floyd *et al.*, 1987), the project developed a participation-oriented method of software development. The main intention of STEPS (see Floyd *et al.*, 1989) was to see a software product result from the process of software development. Accordingly, the use and 'non-use' of software is intertwined with its development: if software is not tailored to the actual needs of the users, they will try to avoid using the software, or loose a lot of time in order to integrate it into their work process. The restructuring of user's work processes is one consequence of the development and implementation of a new software. Therefore, within STEPS the process of software development is the area of primary concern.

STEPS is an evolutionary approach portraying system development in cycles of version production, application and revision. Owing to ongoing social changes (in bureaus, in the factories and in production as well), it is not possible to define all the required functions and the quality of a software system as completely as necessary at any fixed point in time. Therefore the project model has to allow the choice among different strategies and the use of adapted tools.

STEPS is supporting mutual learning of developers and users by establishing and coordinating the processes of cooperation with a minimum of pre-defined intermediate products. These products allow the choice among different ways. Therefore, prototyping techniques are used for experiments and established tools and methods (such as system analysis and design technique (SADT)) were adapted to the needs of cooperation and a more incremental development process.

To increase and manage the coordination, 'reference lines' were used. Reference lines are defined project states to be reached in terms of intermediate products. This is a convenient method for a dynamic coordination of the development process: 'Each reference line is named, and specifies one or more intermediate products, criteria for evaluating them, and procedures for decision-making. The next reference line has to be defined when a thus defined project state is reached, at the latest. Hence, we can prevent from attaining an undefined state' (Floyd *et al.*, 1989: 60).

According to the experiences of the project, the role of a cooperative project establishment has to be highlighted. The initial period of the project has to be used to define a common perspective among the project members. 'In particular, those involved in the project must be prepared to engage in cooperative working processes for developing the system, and agree on a common course of action with its resultant commitments. An inevitable precondition of participative system development is that an agreement on the project's goal and on the way it is going to be achieved has to be worked out, not only between the contractors but especially between the users and developers who are to cooperate together' (Floyd *et al.*, 1989: 59).

Thus, the multiperspectivity of the design process can be used in a productive manner: users, as experts of work process, and software developers, as experts for programming and technical realization, are able to bring their competences together and to discuss options and restrictions of the intended system. Accordingly, STEPS is recurring to a perspective-based evaluation of the software as supreme guide both in building models and in interpreting constructed models in the context of meaningful human activities.

The approach has been developed within a project in which an archive and documentation software has been developed together with the employees using it. The conditions of the project have been fixed in a contract between the workers' council, the management of the institute where the software has been used, and the developers.

What has to be discussed further is whether it is necessary to use to a dynamic perspective of system design and development in other production systems as well: system design is not finished when the first release of the system is running. Due to changing market conditions and changing social objectives within the firm, it is evidently necessary to implement a continuing increase of production. Approaches like STEPS might be able to integrate this in a methodological way.

'Mikropolitik': discussing the reasons of implementation failures

One of the main focuses of the SoTech programme was conducting investigations into the implementation processes of ICTs. Due to the fact, that computer technologies evidently led to a variety of conflicts and problems inside the firms, it was necessary to acquire more detailed information on these changes. The process analysis was directed to the decisions on, and first uses of, ICTs in the firms. The basic considerations of these studies were:

1. The pursuit of individual and egoistic interests has dramatic influence of the implementation of ICTs and new organizational structures inside the firms. For the 'outcome' of these changes, it might be one of the most relevant factors.
2. Due to that, the outcome of an organizational change is not determined in advance by technological choice, for example. There is a contingency of the development.
3. There is no dominant economic rationale, but rationality is limited. Structures (i.e. organizational rules and resources) are the limiting factors for the organizational trajectories. The focus of the investigations is analysing power within organizations.

Ortmann *et al.* (1990) described the implementation processes as ongoing in an 'area of struggle' (*umkämpftes Terrain*): individual and group interests (following divergent rationales) within the firm lead to a redefinition of influence and power spheres inside the firm during the decision and implementation period. These redefinitions and conflicts are regarded as relevant factors determining the outcome and configuration of the system. Every process of shaping technologies is intentionally changing social structures. Firms are social systems in which hierarchy and power structures on the one hand and cooperation and negotiation on the other are existing at the same time (Pries *et al.*, 1990: 212). An area of conflict is thus defined and all divisions and subdivisions involved will try to increase their individual position. This is leading to changing structures and changing objectives as a problem of implementation processes.

This is one reason (apart from explicit decisions) why managements' concepts of implementation of ICTs are modified or broken into pieces and why the objectives change during realization: the companies' objectives are thus becoming moving targets. Ortmann *et al.* (1990: 59ff.) uses the game-metaphor to get an integrated view on these individual objectives, options, activities and changing structures. The regular working processes and production are interpreted as 'routine games' in which cost and benefits are defined and the solving of routine tasks is honoured. 'Innovation games' are processes leading to a new fixing of tasks, influence, and competence during ICTs implementation. For the firms it is a strategic problem to manage a combination of both games in a way that the necessary change and improvement can be realized while the disturbance and turmoil is restricted.

This description of 'micropolitical' processes is another argument for a dynamic definition of process aspects and 'intermediate' products to be taken into account when discussing methodological prerequisites of symbiotic approaches. In trying to come to a defined state at the beginning of the change process inside the firm, it is evidently necessary to clarify the conditions and factors to be fixed and those to be changed during the contracting. 'Mikropolitik' as a theoretical focus of describing change processes is limited. Up to the present, a lot of interesting and fruitful investigations on implementation processes of ICTs in firms have been published (e.g. Birke, 1992). But the step from case studies to broader evidence or to a theory of shaping processes has still to be taken.

Aspects to be investigated further

The present situation in the German discussion on symbiotic approaches is difficult. Due to cutbacks in funding research projects and due to a concentration on problems of healthiness and safety of working life, the federal AuT programme is in danger of losing its initiating position. The ongoing projects of the SoTech programme will be finished due to the decision to end the programme in 1994 and most of the other work-oriented technology programmes of the federal states are unable to fill the shortfall (see Fricke, 1994). Hence, the research infrastructure in Germany for the questions discussed is facing a severe cutback.

On the other hand, semi-autonomous group work and lean production approaches, especially new production concepts, receive public attention as never before. The present situation is characterized by a concordance of the means to be taken in order to solve the productivity problems in German industry. Relevant parts of the unions, the employers' associations and governmental institutions consider an organizational structure based on semi-autonomous groups, for example, to be a fruitful approach. But even on a plant level, the concepts pursued differ significantly and the concordance is proved to be based on the 'wording' in many cases. According to this situation, a new orientation for present research might be in an investigation, and a more active role, of the researchers in change processes with a stronger orientation to an industrial funding of consulting processes. The intention could be threefold (see Brödner and Pekruhl, 1991):

1. As demonstrated in research experiences, the main focus for the changes should be an organizational one. Organizational changes are to be fostered, machinery or technology is to be seen as a means to support working processes and workers' competences, without neglecting the technical aspects and preconditions.
2. The change processes should be organized in participational way. Workers (blue- as well as white-collar) are to be seen as experts of working processes. The knowledge and competences have to be developed and they have to be involved in changing and developing the new work organization.
3. Different sciences (social sciences as well as engineering) have to be integrated in an action-research approach in order to use all available methods for the solution of the firms' problems and an increase of working conditions. The project partners which are able to support this kind of an integrational process, have to be carefully selected. A dissemination of the experiences among scientists, consultants, and practitioners in the firms has to be organized.

According to the aspects discussed, it is still necessary for a further development of symbiotic approaches to focus the process of shaping production systems and organization which is still not adequately structured. Kötter and Volpert (1993) recently criticized the formulation of principles and elements for structuring these processes and that there was no consistent or applicable concept available. As elements of a shaping process they listed:

1. The reference to the individual work task has to be the objective at all stages of shaping the system.
2. The starting point for the (re-)structuring process is a fixing of a common perspective on the situation of the firm at the beginning. This might happen consensually or as documented dissent of the groups involved.
3. It is necessary to define the objectives to be gained during the process; the definition must be prepared in a participational way.
4. A definition of a project team is necessary. The team is aiming at the implementation of the solutions to be developed and it is taking into account the principles of the process. Therefore it is evident that:
   - there are clearly defined competences of the project team inside the firm;
   - the team has to define options and alternatives for the process, timing and operational criteria for measurement of success;
   - a professional moderation and support of the project team must be provided because a neutral (= not company internal) actor is necessary;
   - the shop-floor level has to be involved; the team has to use methods and measures to support this participation.

This list is not complete but it is a step towards a more precise definition of how change processes inside firms can be managed. On the other hand, it is necessary to keep in mind, what cannot be influenced immediately: firm and management culture (shared values, orientations, attitudes and norms) can be developed by using different methods and tools to increase cooperation and communication. In this sense, the obstacles induced by these orientations, the 'soft side' of organization, seem to be the 'harder' category of problems to be solved. But firm culture cannot be shaped. What can be shaped is the organization of production, the work structure, and technical aspects of the production system. But the process to initiate the development of shared visions inside the company can be defined and structured. This is at least one of the research problems and conceptual tasks for a further investigation into symbiotic approaches.

References

von Alemann, U. and Schatz, H. (1986) *Mensch und Technik. Grundlagen und Perspektiven einer sozialverträglichen Technikgestaltung*, Opladen: Westdeutscher Verlag.

von Alemann, U., Schatz, H., Simonis, G., Latniak, E., Liesenfeld, J., Loss, U., Stark, B. and Weiß, W. (1992) *Leitbilder sozialverträglicher Technikgestaltung. Ergebnisbericht des Projekttrgers zum NRW Landesprogramm 'Mensch und Technik — Sozialverträgliche Technikgestaltung'*, Opladen: Westdeutscher Verlag.

Behr, M., Heidenreich, M., Schmidt, G. and Graf von Schwerin, H.A. (1991) *Neue Technologien in der Industrieverwaltung*, Opladen: Westdeutscher Verlag.

Binkelmann, P., Braczyk, H.J. and Seltz, R. (1993) *Entwicklung der Gruppenarbeit in Deutschland*, Frankfurt/New York: Campus.

Birke, M. (1992) *Betriebliche Technikgestaltung und Interessenvertretung als Mikropolitik. Fallstudien zum arbeitspolitischen Umbruch*, Wiesbaden: Deutscher Universitätsverlag.

Brödner, P. (1985) *Fabrik 2000. Alternative Entwicklungspfade in die Fabrik der Zukunft*, Berlin: Edition Sigma.

Brödner, P. and Pekruhl, U. (supported by Hennig, J. and Malberg, M.) (1991) *Rückkehr der Arbeit in die Fabrik. Wettbewerbsfähigkeit durch menschenzentrierte Erneuerung kundenorientierter Produktion*, Gelsenkirchen: IAT.

Brödner, P., Simonis, G. and Paul, H. (Eds) (1991) *Arbeitsgestaltung und partizipative Systementwicklung*, Opladen: Leske & Budrich.

Bundestag (1988) *Antwort der Bundesregierung auf eine Große Anfrage der SPD-Fraktion*, Bundestagsdrucksache 11/3780 vom 22.11.1988.

Christmann, B. and Schmidt-Dilcher, J. (1991) *Die Einführung von CAD als Reorganisationsprozeß*, Opladen: Westdeutscher Verlag.

Dankbaar, B. (1987) Social assessment of workplace technology — some experiences with the german program 'Humanization of Work', *Research Policy*, **16**, 337–352.

Dertouzos, M.L., Lester, R.K., Solow, R.M. and the MIT Commission on Industrial Productivity (1989) *Made in America: Regaining the Productive Edge*, Cambridge, MA: MIT Press.

Ducki, A., Niedermeier, R., Pleiss, C., Lüders, E., Leitner, K., Greiner, B. and Volpert, W. (1993) *Büroalltag unter der Lupe. Schwachstellen von Arbeitsbedingungen erkennen und beheben — ein Praxisleitfaden*, Güttingen: Hogrefe.

Dunckel, H., Volpert, W., Zölch, M., Kreutner, U., Pleiss, C. and Hennes, K. (1993) *Kontrastive Aufgabenanalyse im Büro. Der KABA-Leitfaden. Grundlagen und Manual*, Zürich/Stuttgart: Verlag der Fachvereine/Teubner.

FFES (Forschungsinstitut der Friedrich Ebert Stiftung), Fraunhofer Institut für Systemtechnik und Innovationsforschung, Gesellschaft für Arbeitsschutz und Humanisierungsforschung, Praxisschwerpunkt für betriebliches Organisations und Personalwesen an der Fakultät für Soziologie der Universität Bielefeld, and WZB-UVG (Eds) (1982) *Ein Programm und seine Wirkungen: Analyse von Zielen und Aspekten zur Forschung 'Humanisierung des Arbeitslebens'*, Frankfurt/New York: Campus.

Floyd, C., Mehl, W.M., Riesin, F.M., Schmidt, G. and Wolf, G. (1987) *SCANORAMA — Methoden, Konzepte, Realisierungsbedingungen und Ergebnisse von Initiativen alternativer Softwareentwicklung und -gestaltung in Skandinavien*, Düsseldorf: Minister für Arbeit, Gesundheit und Soziales in Nordrhein-Westfalen.

Floyd, C., Riesin, F.M. and Schmidt, G. (1989) STEPS to software development with users', in: Ghezzi, C. and McDermid, J.A. (Eds), *ESEC '89*, Berlin/Heidelberg/New York: Springer, pp. 48–64.

Fricke, W. (Ed.) (1994) *Arbeit und Technik-Programme in Bund und Ländern 1993. Eine sozialwissenschaftliche Bilanz*, Forum Humane Technikgestaltung Heft 9, Bonn: Friedrich Ebert Stiftung.

Hanewinkel, A. (1986) *Auswertung der Ergebnisse des Forschungsprogramms 'Humanisierung des Arbeitslebens'*, SoTech-Werkstattbericht 6, Düsseldorf: Minister für Arbeit, Gesundheit und Soziales in Nordrhein-Westfalen.

HdA-Projektträger (1981) *Das Programm 'Forschungen zur Humanisierung des Arbeitslebens' — Ergebnisse und Erfahrungen arbeitsorientierter Forschung 1974–1980*, Frankfurt/New York: Campus.

Hennig, J. and Pekruhl, U. (1991) 'Widerspruch zwischen Markt- und Produktionsstrategie? Flexible Spezialisierung in der Investitionsgüterindustrie Nordrhein-Westfalens', discussion Paper IAT-PT 02, Gelsenkirchen: IAT.

Jürgenhake, U. and Lichte, R. (1990) *Vom Experiment zur Normalität — neue Arbeitsstrukturen in einem Kaltwalzwerk in der Bewährung*, Dortmund: Sozialforschungsstelle.

Kiehne, R. and Kohl, W. (1988) Das Konzept der Gruppenarbeit in Fertigungsinseln bei Felten & Guilleaume, in: Roth, S. and Kohl, H. (Eds), *Perspektive Gruppenarbeit*, Köln: Bund Verlag, pp. 187–204.

Kleinschmidt, M. and Pekruhl, U. (1994) *Kooperative Arbeitsstrukturen und Gruppenarbeit in Deutschland. Ergebnisse einer repräsentativen Beschäftigtenbefragung*, IAT Strukturberichterstattung 01, Gelsenkirchen: IAT.

Koschnitzke, T. and Köstner, U. (1989) Beispiel Felten & Guilleaume GmbH, Nordenham, in: Dybowski, G., Herzer, H. and Sonntag, K. (Eds), *Strategien qualitativer Personal- und Bildungsplanung bei technisch-organisatorischen Innovationen*, Neuwied/Frankfurt: Kommentator-Verlag, pp. 69–80.

Kötter, W. and Volpert, W. (1993) Arbeitsgestaltung als Arbeitsaufgabe — ein arbeitspsychologischer Beitrag zu einer Theorie der Gestaltung von Arbeit und Technik,

Zeitschrift für Arbeitswissenschaft, **47** (3), 129–140.

Latniak, E. (1995) *Technikgestaltung und regionale Projekte. Eine Auswertung aus steuerungstheoretischer Perspektive*, Wiesbaden: Deutscher Universitätsverlag.

Latniak, E. and Simonis, G. (1994) Socially oriented technoplogy policy in Germany — experiences of a North Rhine-Westphalian programme, in: Aichholzer, G. and Schienstock, G. (Eds), *Technology Policy. Towards an Integration of Social and Ecological Concerns*, Berlin: De Gruyter, pp. 270–299.

Leitner, K., Volpert, W., Greiner, B., Weber, W.G. and Hennes, K. with cooperation of Oesterreich, R., Resch, M. and Krogoll, T. (1987) *Analyse psychischer Belastung in der Arbeit. Das RHIA-Verfahren. Handbuch*, Köln: Verlag TÜV Rheinland.

Lichte, R. and Trültzsch, K.L. (1986) Arbeitsstrukturierung als sozialer Prozeß. Am Beispiel eines Kaltwalzwerkes, in: Fricke, W., Johannson, K., Krahn, K., Kruse, W., Peter, G. and Volkholz, V. (Eds), *Jahrbuch Arbeit und Technik in Nordrhein-Westfalen*, Bonn: Neue Gesellschaft, pp. 65–76.

Lichte, R. and Reppel, R. (1988) Beteiligungsgruppen im Kaltwalzwerk — ein Modell?, in: Roth, S. and Kohl, H. (Eds), *Perspektive Gruppenarbeit*, Köln: Bund Verlag, pp. 123–137.

Lichte, R. and Jürgenhake, R. (1992) Stahlarbeit: Umbruch in einer konservativen Branche, *Arbeit*, **1** (1), 93–102.

Manz, Th. (1990) *Innovationsprozesse in Klein- und Mittelunternehmen*, Opladen: Westdeutscher Verlag.

Oesterreich, R. and Volpert, W. (1986) Task analysis for work design on the basis of action regulation theory, *Economic and Industrial Democracy*, **7** (4), 503–527.

Ortmann, G., Windeler, A., Becker, A. and Schulz, H.J. (1990) *Computer und Macht in Organisationen*, Opladen: Westdeutscher Verlag.

Ostendorf, B. and Schmid, J. (1994) *Der Einfluß der Betriebsgröße. Analysen und Interpretationen der zweiten Welle des NIFA-Panels*, Arbeitspapier des Sonderforschungsbereichs 187, Bochum: RuhrUniversität.

Peter, G. (1990) 'Implementationserfahrungen mit Umsetzungskonzepten im Rahmen des Humanisierungsprogramms', discussion-paper IAT-PS 06, Gelsenkirchen: IAT.

Pieper, A. and Strötgen, J. (1990) *Produktive Arbeitsorganisation. Handbuch für die Betriebspraxis*, Köln: Deutscher Instituts-Verlag.

Pries, L., Schmidt, R. and Trinczek, R. (1990) *Entwicklungspfade von Industriearbeit*, Opladen: Westdeutscher Verlag.

REFA — Verband für Arbeitsstudien und Betriebsorganisation e.V. (1987) *Methodenlehre der Betriebsorganisation. Planung und Gestaltung komplexer Produktionssysteme*, München: Hanser.

REFA — Verband für Arbeitsstudien und Betriebsorganisation e.V. (1991a) *Methodenlehre der Betriebsorganisation. Grundlagen der Arbeitsgestaltung*, München: Hanser.

REFA — Verband für Arbeitsstudien und Betriebsorganisation e.V. (1991b) *Methodenlehre der Betriebsorganisation. Arbeitsgestaltung im Bürobereich*, München: Hanser.

REFA — Verband für Arbeitsstudien und Betriebsorganisation e.V. (1991c) *Methodenlehre der Betriebsorganisation. Arbeitsgestaltung in der Produktion*, München: Hanser.

Rolf, A., Berger, P., Klischewski, R., Kühn, M., Maßen, A. and Winter, R. (1990) *Technikleitbilder und Büroarbeit. Zwischen Werkzeugperspektive und globalen Vernetzungen*, Opladen: Westdeutscher Verlag.

Roth, S. and Kohl, H. (Eds) (1988) *Perspektive Gruppenarbeit*, Köln: Bund Verlag.

Volkholz, V. (1991) *HdA-Bilanzierung*, Dortmund: Montania.

Volpert, W., Oesterreich, R., Gablenz-Kolakovic, S., Krogoll, T. and Resch, M. (1983) *Verfahren zur Ermittlung von Regulationserfordernissen in der Arbeitstätigkeit (VERA): Analyse von Planungs- und Denkprozessen in der industriellen Produktion*, Köln: TÜV Rheinland.

Womack, J.P., Jones, D.T. and Roos, D. (1990) *The Machine that Changed the World*, New York: Rawson Associates.

6

Managing Sociotechnical Change: A Configuration Approach to Technology Implementation

Richard Badham

Introduction

Since the 1970s, advocates of sociotechnical systems that effectively combine technology, organization and people have been able to point to an increasing number of successful examples of sociotechnical change and both organizational techniques and technological artefacts that support sociotechnical systems. This has been particularly noticeable following the trend towards developing 'high performance' organizations to address rapidly changing and volatile markets (Buchanan and McCalman, 1989), the ongoing contribution of European programmes promoting new production concepts of this kind (Badham and Naschold, 1994), and the effectiveness of Japanese 'lean production' techniques (Womack *et al.*, 1990).

At the same time, the dissemination of sociotechnical systems that effectively delegate responsibility onto self-directed teams and support this through appropriate technological change has not been widespread. Evidence from reviews of European experiences reveal a frequent collapse of many experiments of this kind (Badham and Naschold, 1994). There are a number of contributing factors to this problem: the scarcity of managerial time for undertaking the time-intensive cultural change required; the entrenchment of a traditional engineering mentality amongst engineers and information staff, supported by the educational system; industrial relations systems and cultures that create institutional and attitudinal obstacles to change in this direction; and the frequent failure to understand or address the relative balance of costs and benefits of such systems in different contexts. Given the central importance of addressing these difficulties in sociotechnical projects, one of the key failures of traditional sociotechnical approaches, is the absence of clearly defined methodologies or techniques for managing the change process. Change management, in this context, is not only one of adjusting the organizational structure, culture and technology to the 'demands' of sociotechnical systems, but also of adapting, customizing and developing sociotechnical configurations to fit the context within which it is to operate. Generic techniques for this process of change and configuration, based on, and informed by, case-study experience, are severely lacking. This point is made by a number of contributors in this volume. The development and illustration of an approach to guide this process is the main purpose of this chapter.

The chapter proceeds in three sections. First, a general introduction and critique of traditional sociotechnical approaches. Second, an overview of a new approach to sociotechnical change management: the configurational process approach. Third, a brief introduction to a case study of sociotechnical change that is both informing and being informed by a developing configurational process approach. The case study is an Australian-German project to design and implement team-based manufacturing cells within three Australian companies. The purpose of the project is to provide theoretical and practical insights of sociotechnical change management, informed by the experience of action research in the three firms.

Traditional and modern sociotechnical approaches

Sociotechnical production systems are those in which the interdependent and interpenetrating nature of technology and organization are recognized, and both technology and organization are designed and implemented to support each other. For a number of contributors in this volume, the term 'symbiotic' approach is used to refer to the integration of technology and organization in relationships of

- inextricable interdependence;
- interpenetration;
- mutual support.

The use of this generic term helps to direct attention towards the existence of a general approach to production-system design that incorporates and transcends the variety of different insights provided by advocates of human-centred systems, skill-based automation, anthropocentric systems, work-oriented design, cognitive engineering, computer- and human-integrated manufacturing, as well as modern versions of sociotechnical theory. The term sociotechnical is also used here, however, because of the more widespread familiarity with the concept.

Sociotechnical analysis incorporates a variety of different initiatives aimed at integrating technical and organizational criteria into the design and implementation of process innovations or new production systems. Some useful reviews have been provided of initiatives in this area in both Europe and the USA (Clegg and Symon, 1989; Badham, 1991; Badham and Schallock, 1991; Salzman, 1991; Badham, 1992; van Eijnatten, 1993). For the remainder of this chapter, sociotechnical will be used as an umbrella term for the general perspective that underlies all of these approaches. From the original Tavistock studies during the 1950s to the more recent anthropocentric systems projects of the European Commission, the common focus is upon the effective integration of technology and organization in systems that address the demands of their environment. The general approaches concentrate a great deal on the critique of traditional 'Tayloristic' or 'Technocentric' approaches that do not recognize the need for effective mutually adaptive technical and organizational design (Brödner, 1991). Systems designed using such traditional approaches frequently culminate in sub-optimal results, and are unable to build in flexibility and innovation. This was most clearly apparent in the failure of models of the computer-integrated manufacturing 'factory of the future' prominent during much of the 1980s, and has been widely observed in US reviews of technology implementation (Ebel, 1991; Majchrzak and Gasser, 1991).

Despite such experiences, engineering practices and philosophies often continue to assume that the human being is a source of unreliability and error, to be designed out of new technological systems or strictly controlled through tight technical and organizational specifications and mechanisms. Engineering systems tend to be designed prior to and in isolation from any consideration of the organizational features of new production systems and their technical and human consequences (Badham, 1992; Lund *et al.* 1993). These assumptions are often explicitly disavowed by human-relations practitioners but frequently remain embedded in engineering practices of system design. In reaction, the focus of attention of many sociotechnical projects and publications, consequently, has been to reassert the centrality of the human element in effective system design and implementation, hence the focus on 'human'-centred, 'anthropo'-centric, and 'skill-based' design.

As an action approach, the predominant focus of the more recent technologically oriented sociotechnical approaches has been on the design and implementation of new technical systems, and the importance of integrating human and organizational criteria into this process. The social criteria are not particularly novel, and draw strongly from traditional sociotechnical theory, psychological job design philosophies, and social-psychological visions of group or team work. The novelty lies in the degree to which such criteria are applied in a systematic manner at each stage of system design and implementation. In Europe, a great deal of attention has been paid to the creation of exemplary models of sociotechnical design and implementation — both in the shape of new forms of software and equipment and in implemented working techno-organizational production systems.

Emphasis has also been placed on the need to establish new processes of design and implementation — in contrast to traditional technology led, sequential and hierarchical methods. So far this has concentrated strongly on formal design and participation methods and approaches (Clegg *et al.*, 1989; Corbett *et al.*, 1991). It is extending into a broader concern with the organizational politics of technical and organizational change, however, and with the manner in which sociotechnical design and implementation must address the issues these raise (Buchanan and Boddy, 1992). In contrast to more traditional management approaches, however, sociotechnical approaches have been strongly interdisciplinary in character, integrating the insights of human factors engineering, psychology, engineering, and computer science, with industrial sociology and organizational studies.

More recently, there has been an increased concern with structural and cultural barriers to the use and diffusion of systems designed by sociotechnical methods systems, and the types of change management processes required to achieve success (Charles and Wobbe, 1990–1991; Badham, 1993a). Sociotechnical approaches need to be extended beyond an overly dominant focus on 'incorporating the human factor' in technical design to include a new concern. This concern is with promoting change in techno-organizational configurations, recognizing that these configurations are infused at their very core with power relations, organizational conflict and cultural discord. In the case study outlined below, the project team is explicitly acting as change agents in the transition to new 'production island' structures, and conducting ethnographic research on the politics of the change process and the role of change management techniques in facilitating the promotion of sociotechnical systems. The focus on 'production islands' has been deliberately chosen as these have become the key exemplar within Europe of 'human-centred' production systems. In the case study, the term 'team-based cellular manufacturing' has been used to refer to the same thing: the segmentation of production into product-based cells run by semi-autonomous work teams that take on a growing range and depth of indirect tasks.

In addition, the use of sharp contrasts between 'technology-centred' and 'human-centred' approaches have been qualified in two ways. First, by a recognition of different types of non-technology centred approaches (e.g. the difference between European sociotechnical and Japanese lean production approaches to teamwork; see van Bijsterveld and Huijgen, Chapter 3, and Karlsson, Chapter 4). Second, by a broader focus on the effective integration of technology, people and organization (in systems appropriate to different environmental contexts, Kidd, 1990), rather than a one-dimensional focus on flexible people-centred systems in a generalized context of environmental turbulence and uncertainty (e.g. reduced lead times, greater product variety, shorter product life cycles). This issue is a second major theme in the case study discussed in detail below. Production islands, or team-based manufacturing cells, are being introduced in relatively routine and standardized production environments within Australia, and, in comparison with Germany and Scandinavia, there are less legislative guarantees promoting worker participation and protecting working conditions. In this context, the nature of the new forms of teamwork, the degree to which it approximates to 'human-centred' ideals, and the factors determining its final outcome are the subject of the ethnographic research as well as being an action plan for the change agents.

The European sociotechnical approaches have been strongly affected by a decade of concern about competing with new Japanese manufacturing techniques. Despite initial attempts to define the relations between European symbiotic and Japanese 'lean production' techniques (Charles and Wobbe, 1990–1991), there is still considerable uncertainty and debate about models of 'best practice' production. In this context, the philosophy and techniques of sociotechnical approaches to production system design and implementation are likely to continue to increase in significance and receive a growing degree of attention.

The purpose of the next two sections of this chapter is to discuss in some detail a new approach to sociotechnical change management that both informs and is informed by a case study of sociotechnical design and implementation: that of the Australian-German Smart Manufacturing Techniques: Team Based Cellular Manufacturing project. This case study illustrates not only the nature of sociotechnical approaches but also the current concerns that are informing leading-edge sociotechnical projects. The next section outlines the general configurational process approach, and the subsequent section introduces the case study and initial findings.

Configurational process approach

Background

Traditional approaches to technology implementation are characterized by two fundamental weaknesses. First, 'technology-centred' approaches that neglect human and organizational factors in the implementation process have resulted in the widespread experience of over-budget, over-time and failed technology projects (Majchrzak and Gasser, 1991). Moreover, projects that concentrate solely on the technological aspects of innovation neglect the key importance of integrating technical and organizational factors into high-performance production systems. Second, innovation is frequently identified with the take up and use by organizations of new generic technologies and techniques. As management fashions change, advocates of innovation have shifted attention from computer-integrated manufacturing (1970s), to total quality management (1980s), to business process re-engineering (BPR) (1990s). Whatever the innovation on offer, however, there often tends to be a statement by suppliers and a belief by potential users that the adoption of the generic technique will in itself generate the benefits claimed. Yet this ignores the fact that the majority of the gains from new technology are the result of incremental innovation that occurs after the initial adoption (Rosenberg, 1976). Many of the economic and social benefits to be obtained from new technologies and techniques result from post-adoption configuration of the system to fit the particular technical and organizational context of the user organization (Badham, 1993b). The degree to which this is the case varies considerably, as some technologies and techniques are clearly more established, generic and transportable than others. However, the more complex and broad-ranging the system, and the more it plays the role of embodied organizational practice and control, the more important is the sociotechnical configuration process.

The case study outlined in the next section began with a realization that in order to implement new technology, it was necessary both to address the human and organizational factors as a key component of the project, and to ensure that the process of configuration was effectively carried out. Moreover, it was recognized that traditional human relations, ergonomic and sociotechnical models of technical and organizational change have tended to be insufficient as a guide in the change process. Human factors approaches have tended to view the change process in terms of a conflict between 'Tayloristic' and 'Human Oriented' paradigms, and the need to show the greater efficiency and moral superiority of human-oriented approaches through scientific analysis and demonstration examples. They have consequently failed to adequately grasp the complex and negotiated character of sociotechnical change (Badham, 1991). Ergonomists or human factors engineers have traditionally failed to address the influence upon their activities of their organizational location and consequent resources and authority (Perrow, 1983), and traditional sociotechnical methodologies have been criticized for their naïve neglect of managerial politics and the important and political nature of embedded payment systems (Kelly, 1982). A review of current literature on self-directed work teams also revealed a lack of attention to details of the change process or a tendency to adopt unilinear stage models of change (Caruna, 1993). In order to guide the process of intervention and change, and assist the interpretation and analysis of the project, the project team found it necessary to develop a new model of techno-organizational change. The result is what is described as the 'configurational process' model.

The configurational process model is based on findings that technologies in organizations should be regarded more as malleable and negotiated configurations than inflexible objective systems. These findings are based on research results from five different sources: the 'engineering system' approach of the New Technology Research Group at Southampton University; the 'configurational' approach of James Fleck and the Edinburgh Research Centre on Social Studies of Technology; the configurational process approach of the Centre for Technology and Social Change at Wollongong University; the 'unruly technology' approach of Brian Wynne and the Centre for Science Studies and Social Policy at Lancaster University; and the 'thoroughgoing interpretivism' of Woolgar and Grint at Brunel University Centre for Research on Innovation, Culture and Technology. The approach adopted is that technologies are socially shaped both prior to and after their introduction within organizations; that technologies within organizations are part of complex and changing technical systems; that organizations' technical systems are looser and less rule-governed than is often assumed; and that the nature and requirements of technologies are necessarily interpreted within and influenced by the social context in which they are used (Badham, 1993b).

As clearly identified in sociotechnical analysis, the key component of all production systems is not technology but the 'tasks' carried out by operators using specific technical knowledge, equipment and procedures in the transformation of raw materials into products. Unlike the traditional sociotechnical model, however, the definitions of the social as well as the technical components are not simply viewed in systemic terms but, rather, as looser, more ambiguous, conflictual, mal-integrated and negotiated. The operator, like the technology, is him/herself also 'configured' by interpretations and forces internal and external to the process of production. A key element of the configurational process model is, consequently, the specific configuration of both technology and operators within production. This analysis draws upon the user configuration micro-level research of Woolgar (1991) and the societal effects macro analyses of Sorge and Streeck (1988).

A central additional component of the configurational process model is the inclusion within the system of the role of configurational 'intrepreneurs', i.e. those line managers, support staff, and senior managers involved in ensuring the continued operation and development of production systems. While traditional sociotechnical theories recognize the importance of 'boundary management', they have paid less attention to the crucially important role played in the day-to-day operations of the production system by configurational intrepreneurs ensuring that production systems do not collapse and that they obtain the resources that they need. Moreover, the role of such intrepreneurs is crucial to any change process, they are the 'champions' or the 'obstructors' of change, and crucial mediators of any externally driven attempt to change production systems. In including these configurational intrepreneurs within the configurational process, and refusing to clearly demarcate the 'technical' from the 'social' components of working configurations, the model has been influenced by the system and actor network models of Hughes, Law, Latour and others (Bijker *et al.*, 1987).

The next section provides a more formalized model of the configurational process approach.

Outline of the approach

Configurational processes are dynamic processes that transform raw materials (or inputs) into products (or outputs). They are made up of an interdependent set of technological configurations; operator configurations and configurational intrepreneurs.

Technology consists of the knowledge, equipment and/or procedures that make up the structured, material, 'technical', 'non-human' elements of a production environment. These may be more universal and generic in character as 'tools' to support varying activities in different contexts or more particular and context specific involving the sum total of techniques employed in a particular context to achieve a practical result. The concept of 'technological configuration' points to the importance of *specific constellations* of knowledge, equipment and procedures, and the loosely systemic, complex and locally constituted character of working technological systems. Generic specific technological elements are, within this view, merely a technological resource that only become working technologies when they are appropriately configured in a context (see Figure 6.1). This involves, for example, putting into place the rules and procedures necessary to operate the technology in the local organizational environment; customizing the technology to fit into the organization's technological system and perform the operations required of it in the specific production context; learning about, exploiting and

GENERIC SYSTEMS	CONFIGURATIONS
Generic Identity Across Instances	**Configurations in User Sites** Wide Range of Patterns
Systematicity Underlying coherence governing relation and integration of components	**Looseness or *Ad hoc*ness** Mutually interacting but not necessarily mutually constraining parts
System Dynamic Inherent logic which strongly structures development over time	**No Clear System Dynamic** No fixed rules for integration and operation
Natural Trajectories Defining path for incremental innovation	**Subject to Contingencies** Context dependent innovation paths

Figure 6.1 Configurations versus generic systems. (Compiled after Fleck, 1993)

developing the capabilities of the system utilizing the technology as given, or by adding extra elements or devices, etc. Finally, it includes the set of meanings or interpretations of the technology and its requirements that, to a degree, 'constitute' the technology in a specific operating environment and undermine any *simple* view of the 'non-human' character of such configurations.

Operator configurations are the local set of operation and control personnel and their skills, attitudes, interests and roles. Despite some drawbacks in the use of the term 'operator', it has been employed here to focus attention onto, first, the degree to which the 'socio' component of the 'socio-technical' configuration is not separated from the 'technical' component (there are clear relations of mutual definition, enablement and constrain existing between the technological and operator configurations) and, second, to indicate the theoretical links to research on the constructionist approach to software currently investigating the deployment of images of operators in 'configuring the user' processes (Woolgar, 1991). The everyday techno-organizational configuration process configures operators through the interaction of expectations and demands imposed on them by the technological configuration, the configurational intrepreneurs, themselves and other operators, with the characteristics that they bring to the configurational process from 'outside'. Unlike mechanistic interpretations of sociotechnical systems, operators are not fixed independent 'social' entities that are integrated with a fixed 'technology' to form a system. Operators, and expectations by and about operators, are both configured by and themselves configure, the other parts of the configurational process. This occurs in the process of the design, selection, implementation and running of the technological configuration and in the training, selection, implementation, and routine operating actions of system operators and new forms of work organization.

Configurational intrepreneurs are included within the configurational process because of the key role played by configurational 'champions' in establishing and operating configurations, managing the 'boundaries' of the configuration (e.g. gaining resources for configuration 'requirements' and identifying, thwarting or satisfying external 'demands' on the system — this is exemplified in the setting of performance measures and arguments with accountants over their measures, defining the system in such a way that organizational resources should be committed to it because it links into strategic objectives, etc. In the case studies examined in more detail below, this was evident in the white goods manufacturer in discussions over the measurement of value added by team-building activities. In the automobile components supplier, it was apparent in early definitions of the project as an 'R&D' project — in order to prevent 'commercial' criteria being applied. Later on in the same firm, the project was redefined as a 'real world' project to enrol a new plant manager. The term 'entrepreneur' was to be employed but, like the term 'operator', it has unfortunate limited connotations, in this case a common perception of acting as a capitalist risk-taker. The term 'intrepreneur' is used here to emphasize the internal organizational role played by such actors, however their scope should not necessarily be reduced to such a role as it may involve crucial interventions in inter-organizational relations and in the activities of other organizations. It is, however, used here in order to direct attention onto the active, uncertain and risky process of ensuring that production processes run smoothly, continue to receive support, and gain resources necessary for further development. The term is equivalent to alternative terms utilized in the 'actor network' approaches of Michel Callon, Bruno Latour and John Law, i.e. to refer to 'system builders' (Hughes), 'heterogeneous engineers' (Law), 'engineer sociologists' or 'Sartrean engineers' who manipulate both technical and social elements, and overcome obstacles in both areas, in order to design and implement working technical systems (Bijker *et al.*, 1987). The term 'intrepreneur' was preferred to 'engineer' because of the restricted technicist interpretations of the latter term, and tendency to glorify professional engineers.

No distinction is drawn in the configurational process perspective between technological and organizational champions or intrepreneurs, both technological and operator configurations of the *configurational process* have to be continually enrolled, resources provided, and re-enrolled in order to ensure the persistence and development of the overall system — this extends, of course, to the 'hands on' configurational intrepreneurs themselves. In the literature on technology and organizational change, there is a common reference to the importance of system 'champions'. This requires some elaboration, however, to include both 'on the ground' configurational intrepreneurs who are involved in the detailed implementation and operation of the system and 'high level' enthusiasts who promote and defend the introduction and development of the system in the upper reaches of the organization in order to secure the necessary resources for the new configuration. In each of the firms, the nature of this couplet, and problems and difficulties in its working, have been at the centre of much of the work. In the automobile supplier company, the roles of a plant manager enthusiast and a new business 'people person' were split by the retirement of the plant manager and the lack of enrolment and information given to the chief production engineer — the result was marginalization despite middle level intervention and support. In the Australian white goods manufacturer, the manufacturing director was a hesitant but supportive high level leader, and the CAD/CAM 'roving' engineer was the more detailed project developer. The manufacturing director later retired leaving a potential problem of an absence of higher level support. However, the decisiveness of the new manufacturing director combined with the forcefulness of the 'roving engineer' may overcome this problem. The supervisor closely involved in the project still continued to remain a potential problem. While being positive about change and the need for increased teamwork, he remained unconvinced of the value of the academic-industry project. Consequently, the project continued to remain fragile as both upper-level support and line-management enthusiasm remained in doubt, and the existing plant manager is now departing. The project is relatively well established and difficult to reverse in total, but the commitment to extending the ideas further throughout the plant is suspect.

An important qualification must be made here to the simple takeover of the intrepreneur concept from the 'actor network' approaches. The 'configurational intrepreneur' is not the only source of dynamism and change in the system, or of 'heroic' planned development of the system. Depending on the nature of the configurational process, the technology or the operator may be more or less active and influential in determining the course of change and development. The more democratic the system, for example, the more the operators will take on the function of configurational intrepreneurs. In fact, if we are to fully overcome superficial technical/social or human/non-human splits, it would be important to talk not of people but of tasks that will inevitably be carried out if the configuration survives or develops. It is important, however, that the configurational intrepreneur is included *within* the configurational process because: first, new systems cannot operate or develop without such tasks being fulfilled; second, static models of sociotechnical systems may suggest that *configurational processes* are a sub-system that may be run, developed or undermined from the 'outside' — to include the configurational intrepreneurial tasks within the system emphasizes the internal dynamics of action of the configuration as a continuing process, and helps prevent the type of élitist assumptions built into many 'expert' sociotechnical approaches that view process change as an externally initiated change of a conservative passive system.

At the micro level, the actions of configurational processes are complex and changing. No adequate understanding of their operation can be gained without an interpretation at this level. However, an exclusive focus on the details of local configuring would fail to identify key macro events and the broader power relations and structures that have an impact on the way in which the configuration develops. Without getting into the finer points of traditional social scientific macro/micro and structure/action debates, it is necessary to emphasize here that the paths taken by configurational processes are influenced by, but also influence, broader social contexts and events. In the case of the technological configuration, this is primarily through the interaction with external sources of knowledge, equipment and procedures supplied by basic and applied research institutions, professional agencies, consultants, diffusion agencies, etc. For the operator configuration, this occurs through the medium of the skills, attitudes, interests and roles of system users. Configurational intrepreneurs also have skills, attitudes, interests and roles, linked to broader associational culture and allegiances and their location within the organization.

The external contexts that both supply these inputs into configurations and are also affected by configurational outputs in all these areas are: for the technological configuration — the basic and applied R&D system, professional associations, diffusion agencies, and consultants; for the operator configuration — labour markets, industrial organization, education and training systems, and work ethics; for the configurational engineer — the entrepreneurial culture and associations, product markets, and the objectives, structure and resources of the organization.

The interaction between configurational processes and their 'context' is not to be understood in any simple unilinear deterministic framework nor is there any simple demarcation between the configuration and its environment. In order for the configurational process to survive and develop, decisions have to be made to adapt to or transform the environment. For example, if adequate skills are not supplied by the education and training system, a configurational intrepreneur may adapt the configuration to lower skills, initiate in-house training, or attempt to get new courses established in local educational institutions — the latter type of strategy clearly extending beyond adapting to the environment. In addition, in the process of configuring technology and people, the configurational process changes its character in a way that necessarily has an impact, however small, on the external context. There is, therefore, no clear or unilinear determinism, and, as with any attempt at system definition, the boundary between the 'system' and the 'environment' is not only a fluid one but is also a political issue as definitions of configuration and environment play a key role in configurational strategies. For example, within the Australian white goods manufacturing, the definition of team-based cellular manufacturing as 'moving machines' or as a 'pilot case' for team-based organizational development in the firm, shifted the definition of what was to be included in the change, and had a direct effect on the level of commitment of resources to the project.

Finally, each of these areas that make up the 'industrial culture' in which configurational processes are embedded, can occur at the level of the firm, the industrial sector or region, the nation, international region, or globally. As shown by Aicholzer's study of 'systemic rationalization' through information technology and organizational change in Austrian firms, an understanding of activities at all these levels is essential in understanding 'local' changes (Aicholzer, 1991). While the impression this gives is one of incredible complexity, it is with just this type of complexity that the effective configurational intrepreneur operates (often at the level of 'gut feel' or 'instinct') in the creation and perpetuation of configurational processes.

The general model is outlined in Figure 6.2.

Case study: Smart manufacturing techniques project

There have traditionally been two types of 'human-centred' project: those implementing a 'human-centred' working system in an organization, and those producing generic 'human-centred' software or techniques to assist in the design or operation of working systems. In the former case, the configurational process analysis applies to the design and implementation of a new working production system or the transformation of an existing system, the focus is therefore on the changing character or establishment of the firm's configurational process producing a specific product for use in the producer firm or for direct sale on the market, including the more or less important role of the external actors in the project. The end result of a configurational process analysis is, therefore, the specification of the final character of a working production system and the factors influencing the form that this has taken. In the latter case, the configurational process analysis applies to the design of a new process technology or technique, either for use by a user firm or consultancy institution, or for general sale on the market. The focus is therefore on the nature of the final product and the character of the design team (read configurational process) that produced this product. The end result of a configurational process analysis is, therefore, the specification of the final character of the generic product and the nature of the configurational process that determined the form that it took.

The case study is complex because it undertakes both sets of activities and so can be looked on from either perspective. The project was initiated as an Australian-German collaborative project to cooperate on the design of team-based cellular manufacturing systems (TBCM) in Australia and Germany. TBCM is a relatively clear and established technique for grouping similar parts to be fabricated or assembled into distinct sets, the creation of a number of machine or assembly operation groupings or cells to produce the different sets, the creation of a work team to run the machines or assembly operations

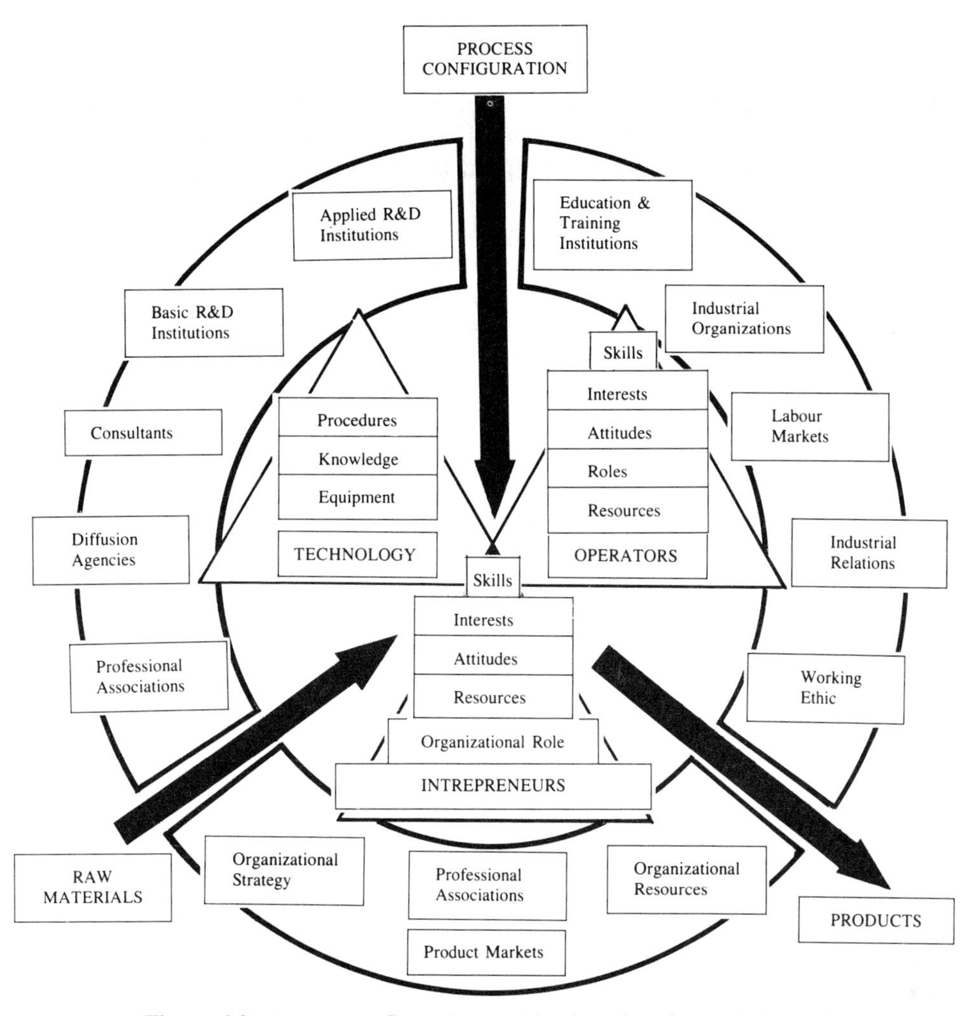

Figure 6.2 Process configuration model. (Compiled from Fleck, 1993)

within each cell, and the devolution of direct and indirect production functions relating to the individual cells (e.g. programming, scheduling, inspection, supervision, set up, maintenance, etc.) to the respective teams. The significance of this technique is its combination and integration of technological and organizational design principles.

The project has been guided by continued Australian-German dialogue over the appropriate nature of TBCM (known in Germany as 'production island'). Within Australia, the project is conducted under the title of the Smart Manufacturing Techniques (SMART): Team-Based Cellular Manufacturing project. This project has been funded for 2 years by the Australian Department of Industry, Technology and Regional Development and three Australian manufacturing companies, and has two aims. First, to design and implement team-based manufacturing cells in a press shop of an Australian white goods manufacturer, in the production and assembly area of an Australian small plastic irrigation products manufacturer, and in the assembly of instrument panels in an Australian automobile supplier company. Second, to accumulate knowledge and experience on techniques for designing and implementing TBCM, and to present these in a form accessible to and able to assist other companies interested in initiating TBCM activities.

Both engineering and social science researchers are involved in the project from the CSIRO Division of Manufacturing Technology, the Engineering department of the University of New South Wales, the South Australian Centre for Manufacturing, and, as project coordinators, the Management of Integrated Technical and Organizational Change (MITOC) centre of the University of Wollongong.

On the one hand the objective of the project is the creation of separate working configurational processes in each of the three firms. Each of these configurational processes can be examined using the model, with 'external' project actors having a more or less important role in the transition from the 'old' to the 'new' configuration. On the other hand the SMART project involves the creation of new software specifications and technical/organizational techniques for team-based cellular manufacturing as 'generic' techniques to be applied to other companies and/or sold in the market.

The complexity of the project and the detailed nature of the configurational process approach make it impossible to adequately describe the overall project in the brief space available here. Moreover, the project, which began in April 1993, is only half completed. Our discussion will, therefore, be restricted to selective observations about the projects from the configurational process sub-projects in the three different firms.

Technological configuration

In each firm, generic TBCM principles and design techniques were insufficient for the design of production and assembly cells in their particular production process. The use of cell layout software to cellularize the white goods manufacturer's press shop was unable to deal with a series of trade offs that compromised the idea of independent cells. For example, decisions about the 'ease' or 'impossibility' of sending different parts through alternative cells, the 'need' to share common machines rather than purchase new machines, change production processes or part designs, etc. These compromises were controversial, often made by engineering groups without broader consultation, and based on informal practical knowledge unavailable to external consultants or other groups in the organization. Moreover, the multitude of details of layout and equipment design were so great that many decisions were made without reference to an explicit and agreed model of how the cells should operate. Day-to-day operational time pressures on the line managers and engineers involved also encouraged this *ad hoc* approach. As a result, the process of data collection to run the cell layout software was a difficult, controversial and political one, influenced in the case of the white goods manufacturer by relatively fixed ideas about the type of cell structure required. Effective engineering design in this context clearly required industrial experience and political acumen. Experienced engineers in cell layout design stress the importance of conducting the analysis on a workbench in the factory in continual 'communication' with the affected parties.

In the automobile components supplier, alternative cell layouts for an instrument panel assembly line were investigated. There was, consequently, considerable controversy over what represented a 'real cell' — for some this was merely a bent line (e.g. 'U-shaped' cell), and alternative forms of mini-lines and parallel lines only fulfilled some team-based cell criteria. The more traditional engineers continued to insist that bent lines constituted cells, while the consultants adopted a view more strongly linked to buffered parallel mini-lines. A formal simulation model was made to assess the costs and benefits of the alternative layouts against agreed design criteria. The analysis pointed clearly to one of the more radical buffered mini-line options. This, however, was circumvented by two events. First, the engineering manager made a decision to bypass the design team and choose a more conservative option on the basis of past firm practices, traditional beliefs and the practical time pressures on the project. Second, the collapse of another assembly line project, meant that the amount of space available for the project was suddenly restricted, thus ruling out most of the options investigated.

In both the white goods manufacturer and the automobile components supplier projects the design of the particular configurational process was also influenced by a variety of factors within the firm but outside the project. The developments within the white goods manufacturer were strongly influenced by: a change from one centralized production scheduling system to another (MRP2 system); a broader company commitment to introducing just-in-time local scheduling; and a change in manufacturing director.

In addition to the equipment, layout and software, the distribution of practical knowledge about the existing configurational process also influenced the ability of different actors to facilitate, influence, obstruct or circumvent the changes being initiated. This was particularly the case when the traditionally oriented manufacturing engineering departments disagreed with the more extreme team-based cellular manufacturing commitments of the research consultants.

A point acknowledged to be of great importance during the project was the nature, degree of formalization, understanding and participation of the cell team in defining formalized work procedures. This formalization process both constrained and enabled shop-floor operators working in cells. As a constraint it imposed certain standard methods of working, and the desirable level of standardization was not determined by TBCM design principles nor had it been openly discussed in the project. With adequate training and experience, however, operators can be trained to understand, modify and develop new procedures.

Operator configuration

Of crucial significance in the design of the configurational process is the level of skills and attitudes of the operators. In both the cases of the white goods manufacturer and the automobile components supplier, most operators were unskilled process workers with no formal post-school qualification and, frequently, low levels of English literacy. At the white goods manufacturer, they were paid very close to the bottom of the industry pay scale, and were one of the worst paid in the urban region in which they were located. Locally, and Australia wide, there was, however, a substantial level of unemployment (over 10 per cent), and, despite the recession, the white goods manufacturer had not laid off any workers. As revealed in weekly discussion groups, the operators were relatively content with their job content, although they revealed a fatalistic resignation to the lack of attention paid to their ideas and needs by middle and senior management. With high levels of unemployment and low skill levels, operators had very little bargaining power. This background created a climate of passivity and cynicism that made it difficult to develop a shared agreement on, and commitment to, the development of the skills and attitudes necessary for the formation of developed cell teams while the final success of team development depends crucially on the nature of the team leader and the degree to which the team becomes dependent on the leader, the background of skills and attitudes acted against such developments. In contrast, however, the plastics irrigation supplier had already gone through a process of general transformation into teams and cells, and is the largest employer in a small rural town with high unemployment. The motivation of the operators to become involved in the design process, as well as their teamwork skills and confidence, were very high.

Different sections of the workforce clearly have different interests in the change. In the white goods manufacturer, for example, for some progress (transport) workers, it means the end of a relaxed job as every operator will also carry out progress tasks, and many expressed passive opposition to the change. For a number of die setters, capable of setting up press machines, it meant a career opportunity to become cell leaders. However, the reluctance of white goods manufacturer management to actively pursue new career paths and payment schemes increased the scepticism and limited the enthusiasm of these members. In another company implementing cell teams in chemical packaging, the skilled trade fitters were described as 'prima donnas' opposed to the change to teamwork.

Broader industrial relations conditions that define the job responsibilities, demarcations and new competency-based career paths are also of crucial significance in determining both the job design framework and training programme and, ultimately, the ability of management and operators to perform the jobs specified by the TBCM design principles. At present, the outcome of industrial relations discussions of competency-based job classifications that are taking place outside the white goods manufacturer, and which will directly impact on the project, have not been resolved.

The interdependence between skill levels and the technical system in the white goods manufacturer were quite clear as most of the presses required continual attention and work by operators. The lack of automatic devices freeing up operators from being tied to machines both reflected and imposed a machine-minder mentality amongst the operators and influenced management attitudes to the operators.

Finally, definitions of the skills and attitudes of the workforce varied between sections and levels in the white goods organization. For some, the press shop was the least skilled and most neglected area of the company, staffed by unmotivated machine minders (e.g. the Personnel Director). For others, the opinion of the press shop operators were seen as possessing a variety of skills which were heavily under-utilized, and they were perceived as having a positive and constructive attitude to the change (e.g. the opinion of the VAM Coordinator). A number of controversial debates over competency standards and the nature and possibility of carrying out questionnaire analyses reflected the political nature of these representations, and the lack of agreement on methods of defining their character.

Configurational intrepreneurs

Much of the literature on sociotechnical redesign, self-directed work teams, and production islands stresses the importance of project 'champions' and 'senior management commitment' to the project. This is rarely theoretically analysed, and there has been no systematic discussion of what type of champions, what level of management, what degree of commitment, and what strategy should be adopted if commitment or champions at particular levels are missing.

In each of the firm projects there has been an important constellation of configurational intrepreneurs. This is a triangle of senior management, line management, and the technostructure. The term technostructure is used in accordance with Mintzberg's definition, i.e. the 'indirect' or 'support' functions involved in determining how operators shall work, through designs for standardization, etc (Mintzberg, 1983). Senior management commitment is required for basic resource commitment, line manager support for the adequate input of informal production knowledge and the active combination of configurational elements in order to bring about the change, and technostructure support is required in order to ensure that senior management commits the necessary resources of time and money on technology and people development, and that line managers do not rush through the project in an *ad hoc* and informal manner that neglects the creation of crucial technical and organizational preconditions.

In each of the projects, the key actors were a senior manager at board level, a plant manager or factory supervisor, and a senior engineer in combination with a human resources manager. The degree of organizational power and attitudes of each of these actors was of crucial significance in defining the character of the project. Where line management was a dominant force, the design and implementation process had an informal but practical and quickly implemented character. Where the technostructure was dominant, the project made more use of formal knowledge and structure but was less able to gain production knowledge quickly or inspire implementation. Where senior management was dominant, the project received adequate funding and resources, but lacked detailed specification and returns were expected very quickly.

In one company, much of the dynamic forces for change came from a senior engineer and a broad ranging but non-managerial human resource person. The lack of full understanding and commitment from senior management and the factory supervisor had direct effects on allocation of adequate time and funding for team development and adequate use of technical and personnel knowledge in the design process. In another company, the departure of the plant manager, resistance from the chief engineer (technostructure), and the lack of influence of the area supervisor, were effective in fundamentally restricting the nature of the cell design to an extremely conservative option.

The external researchers only had an influence on the project to the extent that they could gain commitment from and adequately support members of this triangle. This has made it necessary for the researchers to act as internal coalition builders and change agents. This was of key importance, for example, in activities such as: the effective collection of information for the cell layout system and the use of its recommendations; the establishment of an effective and participatory job design committee; influencing the performance measures group to broaden its productivity measures; promoting visits to other factories and involving shop-floor operators in awareness and design activities; lobbying for the provision of adequate levels of training; ensuring that the project fits into senior management plans and that senior management is educated about the nature and value of the project, etc.

Conclusion

A new era is emerging for technology implementation and sociotechnical system design, inspired by a broader understanding of technological and organizational politics. There now exist a small but growing number of different case studies of the politics of change, and a number of general theoretical schemas are emerging which attempt to conceptualize the process of techno-organizational change. In order to inform research and guide policy, new theoretical approaches are needed that are closely tied into in-depth empirical observations of change. The provision of such a model is the research aim of the SMART project. The configurational process model outlined above has been developed as part of the project and has been found to be useful as a conceptual map and generator of hypotheses. The purpose of this paper has been to introduce and illustrate this model as a contribution to sociotechnical approaches to the change management, implementation studies, and the management of technology. So far, the SMART project has found this framework a highly useful basis for collecting and categorising information and, to a degree, guiding action research.

A major conclusion to emerge from the first year of the SMART project is the central significance of project management and change facilitation in the design and implementation of TBCM. The development and application of appropriate techniques in these areas (strategic/design) appear more important than the creation of either software or group organization techniques that support the effective operation of TBCM once in place. This poses a major challenge to sociotechnical action researchers for it requires them to redirect their focus of attention and systematize what they learn from their experience as agents of change. The extent to which the output of such an exercise can be the delivery of 'techniques' to assist other projects in what is essentially a socio-political process is a matter for debate. It does, however, pose a major challenge for sociotechnical action researchers.

References

Aicholzer, G. (1991) Systemic rationalization in Austria: Social and political mediation in technology use and work organization, *AI and Society*, **5** (4), pp. 277–296.

Badham, R. (1991) Computer integrated manufacturing: The social dimension. A comment, *International Labour Review*, **130** (3), 373–392.

Badham, R. (1992) Skill based automation: Current European approaches and their international relevance, *Prometheus*, **10** (2), 239–259.

Badham, R. (1993a) *Information Technology as Configuration*, MITOC Working Paper 1, Wollongong: University of Wollongong.

Badham, R. (Ed.) (1993b) *International Journal of Human Factors in Manufacturing; Special Issue: Systems, Networks and Configurations; Inside the Implementation Process*, **3** (1).

Badham, R. and Naschold, F. (1994) New technology policy concepts, in Aicholzer, G. and Schienstock, G. (Eds) *Technology Policy: Towards an Integration of Social and Ecological Concerns*, Berlin: De Gruyter, pp. 13–25.

Badham, R. and Schallock, B. (1991) Human Factors in CIM development: A human centred view from Europe, *International Journal of Human Factors in Manufacturing*, **1** (2), 121–141.

Bijker, W.E., Hughes, T.P. and Pinch, T.J. (Eds) (1987) *The Social Construction of Technological Systems: New Directions in the Sociology and History and Technology*, Cambridge, MA: MIT Press.

Brödner, P. (1991) *The Shape of Future Technology: The Anthropocentric Alternative*, London: Springer Verlag.

Buchanan, D. and Boddy, D. (1992) *The Expertise of the Change Agent: Public Performance and Backstage Activity*, Hemel Hempstead: Prentice Hall International.

Buchanan, D.A. and McCalman, J. (1989) *High Performance Work Systems: the Digital Experience*, London: Routledge.

Caruna, L. (1993) 'Self-directed work teams at BHP', unpublished MBA thesis, University of Wollongong.

Charles, T. and Wobbe, W. (Eds) (1990–1991) *APS Research Paper Series*, FAST Occasional papers 245–271, Brussels: Commission of the European Communities.

Clegg, C.W. and Symon, G. (1989) A review of human-centred manufacturing technology and a framework for its design and evaluation, *International Review of Ergonomics*, **2**, 15–47.

Clegg, C.W., Ravden, S., Corbett, J.M. and Johnson, G. (1989) Allocating functions in computer integrated manufacturing: A review and a new method, *Behaviour and Information Technology*, **8**, 175–190.

Corbett, J.M., Rassmussen, L.B. and Rauner, F. (1991) *Crossing the Border: The Social and Engineering Design of Computer Integrated Manufacturing Systems*, London: Springer Verlag.

Ebel, K.-H. (1991) *Computer Integrated Manufacturing: The Social Dimension*, Geneva: International Labour Office.

van Eijnatten, F. (1993) *The Paradigm that Changed the Workplace*, Assen: Van Gorcum.

Fleck, J. (1993), 'Configurations: Crystallizing Contingency', *International Journal of Human Factors in Manufacturing*, **3** (1), 15–36.

Kelly, J.E. (1982) *Scientific Management, Job Redesign and Work Performance*, London: Academic Press.

Kidd, P. (1990) *Organization, People and Technology in European Manufacturing*, FAST occasional paper 247, Brussels: Commission of the European Communities.

Lund, R.T., Bishop, A., Newman, N. and Salzman, H. (1993) *Designed to Work: Production Systems and People*, Englewood Cliffs: Prentice Hall.

Majchrzak, A. and Gasser, L. (1991) On using artificial intelligence to integrate the design of organizational and process change in US manufacturing, *AI and Society*, **5** (4), 321–338.

Mintzberg, H. (1983) *Structure in Fives: Designing Effective Organizations*, Englewood Cliffs: Prentice Hall.

Perrow, C. (1983) The organizational context of human factors engineering, *Administrative Science Quarterly*, **28** (4), 521–541.

Rosenberg, N. (1976) *Perspectives on Technology*, Cambridge: Cambridge University Press.

Salzman, H. (1991) Engineering perspectives and technology design in the United States, *AI and Society*, **5** (4), 339–356.

Sorge, A. and Streeck, W. (1988) Industrial relations and technical change: The case for an extended perspective, in: Hyman, R. and Streeck, W. (Eds), *New Technology and Industrial Relations*, Oxford/New York: Blackwell, pp. 19–47.

Womack, J.P., Jones, D.T. and Roos, D. (1990) *The Machine that Changed the World*, New York: Rawson Associates.

Woolgar, S. (1991) Configuring the user, in Law, J. (Ed.), *A Sociology of Monsters: Essays on Power, Technology and Democracy*, London: Routledge.

7

A Practical Theory and Tool for Specifying Sociotechnical Requirements to Achieve Organizational Effectiveness

Ann Majchrzak and Linda Finley

Twenty-first-century manufacturing

The environment that US manufacturing companies face today and continuing into the twenty-first century is significantly more dynamic than that of yesterday (Smith and Reinertsen, 1991; Winby, 1993). These trends include:

- globalization of technology and markets;
- fragmented, sophisticated, and demanding customers;
- smart products with fused technologies and software as a major component;
- rapid product and process technology changes;
- customers intervening in the design of new products;
- the formation of value-added alliances;
- environmentally conscious manufacturing.

These trends characterize an environment that places exceeding pressure on US manufacturing enterprises to compete not just on cost, but on speed, consistency, acuity, agility, and innovativeness (Stalk *et al.*, 1992). A company that is able to accurately anticipate customers' evolving needs, respond quickly to those needs, adapt simultaneously to many different needs, and consistently produce a satisfying product will be the successful company of the future.

To meet these competitive challenges, the successful company will need to change its approach to work organization and management. Fundamental to this change is increasing the enterprise's ability to integrate the people more completely, and also the technology, organizational, and strategic needs of the enterprise. Integrating these needs means that the systems that manage people, technology, organizational and strategic elements are complementing each other in meeting the competitive challenges, rather than suboptimizing or working in different directions (Senge, 1990). This change is fundamental to at least three initiatives currently popular in the USA: agile manufacturing, business process re-engineering, and concurrent engineering.

In agile manufacturing, the emphasis is on the rapid development and implementation of new product and process technology (Nagel and Dove, 1991). The agile manufacturer is able to get products to a market quicker than the competitors and thus reap the benefits of fast development, such as increased market share, higher profit margins, and lasting learning curve advantages (Hayes *et al.*, 1988; Clark and Fujimoto, 1991; Smith and Reinertsen, 1991). Critical to bringing new product and process technologies to market rapidly is effective planning for integrating the organizational, people, and strategic elements with the needs and capabilities of the technology. For example, a report from the National Research Council's Manufacturing Studies Board, based on 24 cases of the implementation of CAM and CIM technologies, concluded that

> Realizing the full benefits of these technologies will require systematic change in the management of people and machines including planning, plant culture, plant organizations, job design, compensation, selection and training, and labor management relations (Manufacturing Studies Board, 1986: 2).

More recently, researchers concluded that major changes in manufacturing systems, such as those represented by CIM and JIT, have direct implications for the overall organizational design. Thus, it may be more productive to redesign the organizational structure before implementing available technology than to hope the technology will bring about manufacturing effectiveness (Duimering *et al.*, 1993: 55). Major changes in products also have an impact on organizational systems, irrespective of their impacts on the manufacturing systems. New product development studies often conclude that the most successful products are the ones in which planning is emphasized in the front end to account for potential downstream effects to delivery, maintainability, and continuous improvement (Smith and Reinertsen, 1991).

Planning for integration is critical for business process re-engineering (BPR), as well. BPR is the mapping and redesign of process steps to achieve significant costs in cycle times (Hammer, 1990; Hammer and Champy, 1993). For AT&T Global Business Communication systems, as a result of a re-engineered process step, job descriptions for hundreds of people needed to be rewritten, new recognition and reward systems developed, the computer system revamped, massive retraining undertaken, financial reporting extensively revised, and shipping and billing with suppliers and customers rearranged (Stewart, 1993). If these design choices are wrong, BPR can result in an expensive failure.

Finally, for concurrent engineering (CE), planning for integration is critical as well. CE is the practice of having teams of designers represent the complete spectrum of design disciplines and interactively and simultaneously design new products and processes to reduce new product introduction costs and increase agility, quality and serviceability of the product. By definition, CE involves the effective integration of such organizational initiatives as teaming, such people issues as cross-training, such strategic issues as product maintainability, and such technology issues as CAD and CIM and product data exchange standards (Carter and Baker, 1992).

To summarize, for manufacturing enterprises to prepare for the twenty-first-century will involve the effective integration of organizational, people, strategic, and technological issues. However, do we have practical theories and tools to aid enterprises in this preparation?

The STS paradigm

The most useful paradigm for effecting organizational, people and technological integration is the sociotechnical systems (STS) paradigm (Taylor and Felten, 1993; Trist and Murray, 1993). Other chapters in this book describe the STS paradigm in significant detail; therefore, only a brief summary is provided here. The essence of the sociotechnical systems paradigm is that organizational systems function effectively and proactively because all the elements of the system are compatible (or integrated) with each other, and not because of the characteristics of any one element taken alone. A list of elements illustrating the comprehensiveness of the characteristics to consider in an organizational system, is presented in Figure 7.1.

According to the STS paradigm, integration among all of these elements is achieved by considering that any and all elements affect each other and, thus, any changes to one element are likely to create changes to other elements. Therefore, changes to technology are likely to create misalignments with the organizational and people systems; misalignments that create organizational effectiveness until the people and organizational systems are changed. Because of the interdependencies among the elements, the STS approach suggests that all these elements should be considered early in a planning process to ensure that the changes needed are feasible, appropriate and tend to optimize the entire system rather than suboptimize to meet individual needs of individual elements in the system. So, for example, in a business process re-engineering activity, changes to the process map are likely to cause changes in the organizational design and thus both process and organizational changes should be planned simultaneously to ensure that both complement each other when the changes are implemented.

In the USA, STS tends to be practised primarily as a value-laden methodology. The values include: a consideration for continuous improvement, creating a self-empowered team-based culture, democratic design, and the importance of quality of worklife for workers. The methodology of STS, as described by Taylor and Felten (1993) includes:

1. discovery (learning and creating design structure);
2. open system scan;
3. technical and social analyses;
4. design of jointly optimized technical and social analysis;
5. implementation.

Limited diffusion of STS

As described by Taylor and Felten, STS in the USA has diffused to a number of large companies (including, for example, Hewlett-Packard, Digital Equipment Corporation, Proctor and Gamble, and General Foods). However, the diffusion has been limited because of the strong influence of the scientific management methods currently applied to integration planning in the USA. A number of reasons can be cited for this limited diffusion. These reasons include:

- elapsed time that STS takes;
- consultant dependency;
- manual analytic techniques;
- topdown process;
- teams as primary solution;
- the great learning curve or paradigm-shift.

Elapsed time that STS takes

The typical STS intervention often takes participants 18 months from initial scan to design. This amount of time is often needed because participants need to learn the

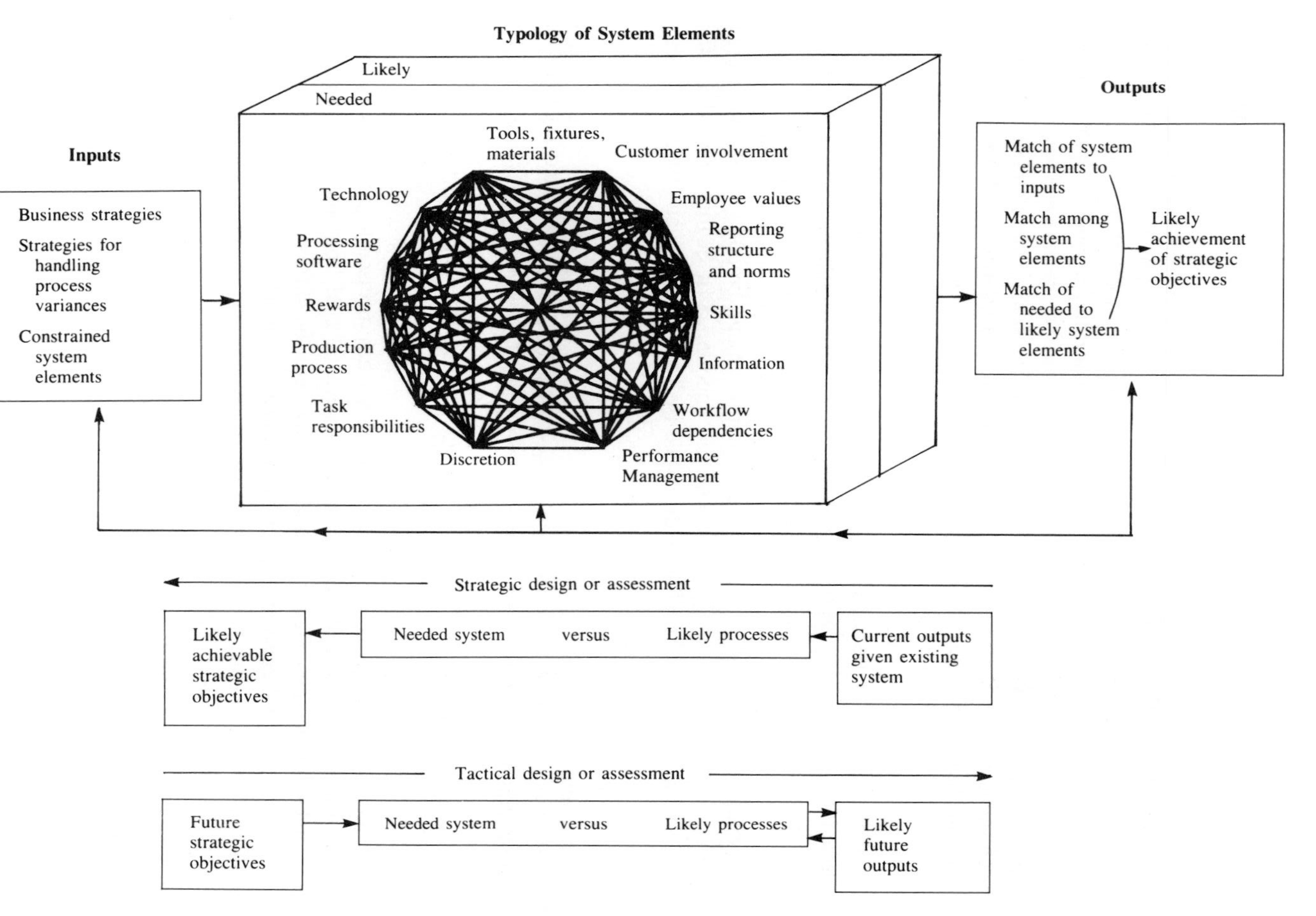

Figure 7.1 ACTION open systems model

language, become accustomed to the STS perspective, obtain the necessary data for analysis, share design ideas with those outside of the design team, converge on a design approach, and communicate that design to others in the organization. Often the participants are working for the first time as a team and typically as a part-time effort.

Consultant dependency

STS, as practised in the USA, is often done in companies only with the help of STS consultants. This dependency on consultants arises from a number of different factors including: paying a consultant helps to maintain focus on the effort, an outside objective person viewing the process can provide valuable assistance at critical junctures in the STS process, and the consultant brings a rich knowledge base of relationships among system elements based on experiences with best practice companies and previous experiences.

Manual analytic techniques

Most analytic techniques used in STS are manual in nature; computer simulations except as they pertain strictly to technical changes such as redesigns of layouts are rarely used. As a result, data collection, analysis and interpretation of the data to create designs is so time-consuming as to discourage detailed analysis of the likely impacts of alternative designs for comparison purposes.

Top-down process

While the values of STS encourage democratic design, the reality is often that a top-down, elitist process is undertaken in which a management steering committee defines the mission and focus of the design team (and often even selects the design-team members), the steering committee reserves the right to approve and disapprove design recommendations, and the design team often becomes perceived as the inner-design group who has more influence on the design process than those not selected to be on the team. Search conferences, a recent innovation in STS methodology, is intended to overcome this latter problem; however, there is still an 'inner' and 'outer' group.

Teams as primary solution

Most STS processes in the USA end with a recommendation for self-managing work teams. This is unfortunate because in many situations, more complex recommendations are needed (such as matrix organizational designs, and multiple and flexible team memberships).

STS necessitates a great learning curve or paradigm-shift

As one manager in an organization expressed it, 'if you don't feel STS in your heart, you can't be sold on it'. Thus, the chicken-and-the-egg problem is that if you have not got it, you cannot get it.

Dilemmas of STS paradigm

We would argue that many of these limitations arise from the way in which several dilemmas of the STS paradigm are currently managed in the USA STS community. These dilemmas are:

- STS concepts are abstract;
- the large number of variables in a sociotechnical system;
- the large number of relationships among the variables;
- trade-offs are continuous and difficult.

STS concepts are abstract

Many STS concepts are abstract. For example, the STS principle that states that process variances should be controlled at their source contains three abstract concepts: process variances, control, and source. The abstract nature of these concepts is in a sense the strength of the STS paradigm because the concepts are generalizable to many different situations. However, the operationalization of these concepts in real organizations is often a difficult and time-consuming process. For example, the concept of process variances is often misunderstood to be problems that people have with the process rather than known and possibly desired variations from a standard (such as customized parts) (Taylor and Felten, 1993). Thus the dilemma is one of understanding and applying these abstract concepts so that they have real meaning in the redesign context. To cope with this dilemma, many design teams must engage in a significant learning curve process of STS training followed by reinforcement and further education by the STS consultant as attempts are made to apply specific operationalizations of the abstract concepts to their business context.

Large number of variables in a sociotechnical system

The number of variables to consider in designing any sociotechnical system is enormous. To name a few variables — skills, tasks, information sources, hardware, software, areas of discretion, performance metrics, and rewards all are part of a sociotechnical system. For each variable, a design decision needs to be made, such as whether or not to provide a skill or the appropriate characteristics of the hardware or user interface. In the STS paradigm, all these variables should be simultaneously designed to ensure joint optimization. The dilemma then is that the simultaneous design of so many variables is impossible, both because there is too much information to keep in one's head and also because people can only manually make one decision at a time. To handle this dilemma in STS today, design teams may decide to reduce the number of variables that they must design by prematurely judging certain variables to be constrained and thus not worthy of design attention. In one organization, the design team decided that the layout could not be altered because process engineers gave an excellent reason for the need for an assembly-line layout, despite the fact that the line-layout seriously undermined many efforts to create a feeling of task significance and task identity. Thus, agreeing to ignore layout in the design decision-making made the design team feel that at least there was one variable that they were no longer responsible for designing. Another method that design teams use to handle this dilemma of too many variables is to design some of the variables early, recognizing that the design of the later variables will be constrained by the design decisions made about the former variables. So, for example, many STS consultants prefer to design the team structure first, such as the tasks allocated to the team, and leave the design of the performance measurement and reward systems to later decision-making, even after the teams have been implemented. By leaving performance measurement decisions until later, the design teams simplify their immediate design problem. However, the choice of what to design earlier and later has significant implications for the outcome of the design process. One could argue, for example, that the performance measurement and reward systems should be designed first so that work units are designed to fit the performance needs of the system.

Large number of relationships among the variables

The STS paradigm states that the relationships among the variables are as important as the variables themselves, and that each variable has some interaction with other variables in the sociotechnical system. Given that the number of variables in an sociotechnical system are large, it is not surprising that the number of relationships becomes enormous. Moreover, these relationships are not all of the same class: some are common sense, some are derivable from empirically-supported theory, some are derived only from experience or knowledge of best-practice companies, some are still not yet resolved, and some are so context-specific that general relationships are not meaningful. Compiling these relationships into one place, such as the design team's collective brain, and accessing these relationships when needed in a design process is an enormous, if not impossible, task. To cope with this dilemma, design teams may either try and compile these relationships through an extensive learning curve (i.e. by documenting their existing organization in enormous detail and conducting extensive site visits with best-practice companies), rely heavily on consultants to provide this knowledge base, or focus on only what they might prematurely consider to be the key relationships (such as the impact of teams on performance reward systems and ignore the impact of layout on collective action).

Trade-offs are continuous and difficult

The relationships describe impacts of one variable on another. Thus, in an ideal design, all these relationships should be optimized. So for example, if a team structure has been implemented, and the relationship indicates that teams function better when members' rewards are based on team performance, then the ideal design is one in which teams receive team rewards. However, the dilemma is that, despite knowing the ideal relationships, there is rarely a situation in which all ideal relationships can be designed. The design team is constantly making trade-off decisions about which ideal relationships they will follow and which they will ignore. Typical STS redesigns rarely document the process the design team went through to decide which relationships essentially carried more weight in the design. For example, when changes to the performance measurement and reward are not designed into the initial STS design, an implicit decision has been made that violating ideal relationships concerning performance measurement and reward systems is somehow less damaging than violating other ideal relationships. A rationale for these differential weights is rarely provided.

ACTION's approach

In summary, we believe that one of the reasons for the limited diffusion of the STS approach in the USA is its inherent complexity. Having large numbers of abstract variables and ideal relationships, each of which must be reinterpreted and weighted in each particular business context is a task far more complex than most people have the time and resources to manage. The problem statement with which we began our research was: 'How do we manage this complexity for the business community so that the STS paradigm can diffuse more broadly?'

One approach to addressing this problem statement would have been to reduce the complexity of the STS paradigm by reducing the number of variables and relationships. We felt that such a reductionist approach was inappropriate at this time because insufficient empirical evidence (such as from sensitivity analyses) had been conducted to support selective attention to only a few system variables and relationships. Instead, our approach to managing the complexity was threefold:

- construct a knowledge base with a relatively comprehensive list of variables and relationships among the variables;
- create software and a user interface to allow easy manipulation of the knowledge base to make sociotechnical design decisions;
- develop a methodology for incorporating the knowledge base and software into the sociotechnical design process.

The resulting tool is called ACTION (recently renamed to TOP — Integrator for Technology, Organization, and People). Its development has been sponsored originally by Digital Equipment Corporation, and then by a consortium of companies (including Texas Instruments, Hughes, Hewlett-Packard, General Motors, and EDS) through the National Center for Manufacturing Sciences (NCMS). The mission of the NCMS is to develop methods and tools to help US competitiveness in large-, small-, and medium-sized discrete parts manufacturers. The tool has been under development for 4 years, and is currently being pilot-tested in the sponsorship companies. The intention is that, once pilot-tested, the tool will be made available to the complete NCMS membership, which includes over 140 companies. This article will only describe the knowledge base and software, and the exemplar use of the tool in one of the sponsoring companies, Texas Instruments. Because of the nature of our sponsors, the knowledge base has initially focused only on discrete parts manufacturing; however, intentions to broaden out the knowledge base to other industries and other functions in an enterprise (such as design, software development and service) are under discussion.

Knowledge base

To construct the knowledge base of STS variables and relationships, five knowledge sources were queried:

1. Review of industry standards such as NCMS's Achieving Manufacturing Excellence programme, IDEF modelling procedures, business process re-engineering procedures and concepts, Baldrige National Quality Award criteria, and agile manufacturing concepts.

2. Review of theoretical and anecdotal literatures to identify propositions, such as literature on manufacturing strategy (e.g. Porter, 1985; Hayes *et al.*, 1988), STS (e.g. Pava, 1983; Eason, 1988; Senge, 1990), teams and work design (e.g. Hackman, 1990), empowerment (e.g. Lawler, 1988), organizational design (e.g. Thompson, 1962; Woodward, 1982), mental and cognitive workload (e.g. Gagne, 1985), performance measurement and reward, and job analysis (e.g. McCormick, 1976; Champion, 1983; Gael, 1984).
3. Qualitative meta-analysis of empirical studies on implementation of manufacturing technology (see, for example, Majchrzak, 1988; Liker *et al.*, 1993).
4. A 2 $\frac{1}{2}$-year long series of consensus-building meetings with an industry team in which each specific relationship was discussed in detail and agreed to by industrial engineering experts from different discrete-parts manufacturing industries (including electronics, automobile, and metalworking).
5. Three-day case visits with 90 US plants using a pilot-tested, standardized interview and observation protocol collecting information on over 500 sociotechnical variables. These data were used to first confirm the comprehensiveness of the lists of STS variables, and then to confirm the relationships among the variables. (The latter activity of confirming relationships among variables is still continuing.)

As a result of these efforts, a knowledge base was constructed. There are three aspects of the knowledge base. The knowledge base consists first of a reasonably comprehensive list of operational features describing a sociotechnical work system within a discrete parts manufacturer. The list is provided in Table 7.1.

For example, there were seven business objectives or purposes which the work unit could be trying to achieve, any of which could be selected by a socio-technical designer as important to the work unit. Similarly, a list of 18 process variances was developed, any of which could be selected by the ACTION user as relevant and important to the work unit being designed. This list is meant to be sufficiently abstract to be usable in a range of discrete parts manufacturing settings, and sufficiently concrete to provide immediate application to each company's business environment.

Second, the knowledge base consists of a large number of the ideal relationships among the operational features. Hypothetically, the knowledge base could contain ideal relationships among specific features in each of the system elements portrayed in the centre of Figure 7.1, and between each of these system elements and each of the inputs in Figure 7.1.

Given that some of the processes can contain up to 141 specific features, the number of ideal relationships among each would be enormous to specify. Thus, we explored ways of reducing the space of possible relationships. The most obvious way was to make the assumption of transitivity, i.e., that if Variable X and Variable Y are ideally related in a certain way, and Variable Y and Variable Z are ideally related in a certain way, then an inference about the ideal relationship between Variable X and Variable Z could be inferred without explicitly stating it. For this assumption of transitivity to be correct, the relationships between Variables X and Y and between Variables Y and Z would need to be sufficiently encompassing so as to account for the relationship between Variables X and Z. Thus, the potential space of relationships portrayed in Figure 7.1 was reviewed to identify those encompassing relationships. This review yielded three types of relationships as the most encompassing:

- relationships between each business objectives and each operational feature;

Table 7.1 List of sociotechnical features in ACTION

Category of STS variables	Number of variables in category	Example of variable
Business objectives	8	Degree to which all selected goals are facilitated by the choice of other goals
Activities	141	Process design and programming
Norms	6	Norms for collaboration
Customer involvement	5	Is customer directly and frequently involved in long-term scheduling?
Tools, fixtures, materials	3	Tools
Software	10	Part preparation software
Information	22	Tool status and performance information
Technology	20	Part design/process program creation system
Employee values	7	Working closely with others
Production contexts	4	Transfer line
Process variances	19	Ability to run parts that are not standard
Reporting structure characteristics	5	Teams exist where tasks are at least partially shared among team members
Skills	30	Reading
Performance management and reward	15	Performance standards exist?
Discretion	9	Discretion over how work is performed
Production process characteristics	16	Variation in tools, batch, and materials
Specific attributes of skills, technology, information, tools, fixtures and materials	33	Is technology of low complexity?

- relationships between strategy for coping with each process variance and each operational feature;
- relationships between tasks and each operational feature.

We propose that these three sets of ideal relationships account for a large enough possible space of relationships such that other relationships can be inferred. For example, the ideal relationship between a business objective (Variable X) and a technology states that the human override (Variable Y) of the technology is needed to meet the business objective ($X \leftrightarrow Y$). Moreover, the ideal relationships between business objective (Variable X) and discretion states that people should have discretion (Variable Z) over task sequencing ($X \leftrightarrow Z$). The fact that human override (Variable Y) and discretion (Variable Z) are ideally related ($Y \leftrightarrow Z$) can be inferred from the previous two relationships without explicitly stating.

Having identified a reasonable set of possible ideal relationships, the five knowledge sources identified above were reviewed to define the nature of those interrelationships. Approximately 17 000 ideal relationships were identified and approved by the industry team over a 1 $\frac{1}{2}$ year development period.

The third feature of the knowledge base is what we have come to call the sensitivity analysis model. As real life poses many constraints on redesign efforts, many of the ideal relationships will not be feasibly designed into a real implementable design. As a result, design teams must often make trade-offs about which ideal relationships will take precedence over other ideal relationships. To help design teams make these trade-offs, we developed a theory for integrating the features into a single model for predicting organizational effectiveness. The model, based primarily on organizational design and STS literatures, states that organizational effectiveness is primarily a function of appropriately designing three aspects of an organization:

1. differentiation (i.e. designing motivationally complete work roles that support the purpose of the organization);
2. integration (i.e. designing mechanisms for work roles to coordinate with others to obtain needed resources);
3. feedback and control systems (i.e. designing performance measurement and reward systems so that role holders are motivated and focuses on performing in ways that support the purpose of the organization).

The model states that the more that each aspect has been appropriately designed, the greater the likelihood of achieving organizational effectiveness. The appropriate design of each aspect is a function of the full range of sociotechnical systems features. So, for example, the appropriate design of work roles is a function of the resources, skills, customer involvement, tasks, value-biases, discretion, and information provided to the work role definition. The proposition is that the more that these features match (i.e. meet the ideal relationships in the knowledge base) to the particular needs specified by the organization (as defined by its selected business objectives and strategies for coping with process variances), the more likely that aspect is appropriately designed. The model is depicted in Figure 7.2.

In providing a model like that presented in Figure 7.2, assumptions need to be made about the relative criticality of the various design features. For example, is having the appropriate differentiation as important as having the appropriate integration or, for example, cannot having the needed resources allocated to a work role be overcome by good integration mechanisms? We have made the following assumptions in the model:

- Generally, achieving appropriate designs for each aspect are equally important in determining organizational effectiveness; i.e. no other aspect is more important than the other.
- All aspects can be compensated to some degree by the design of other aspects provided that each aspect has at least some of the correct design features; that is, where an aspect is not designed perfectly, the features inappropriately designed can often be compensated by the design of other aspects. For example, not having the complete set of skills needed can be compensated by having experts outside the work role available to provide those needed skills. This means then that no aspect needs to be perfectly designed. For example, a work role can function sufficiently provided it has people with most of the skills needed, but not necessarily with all of the skills needed.

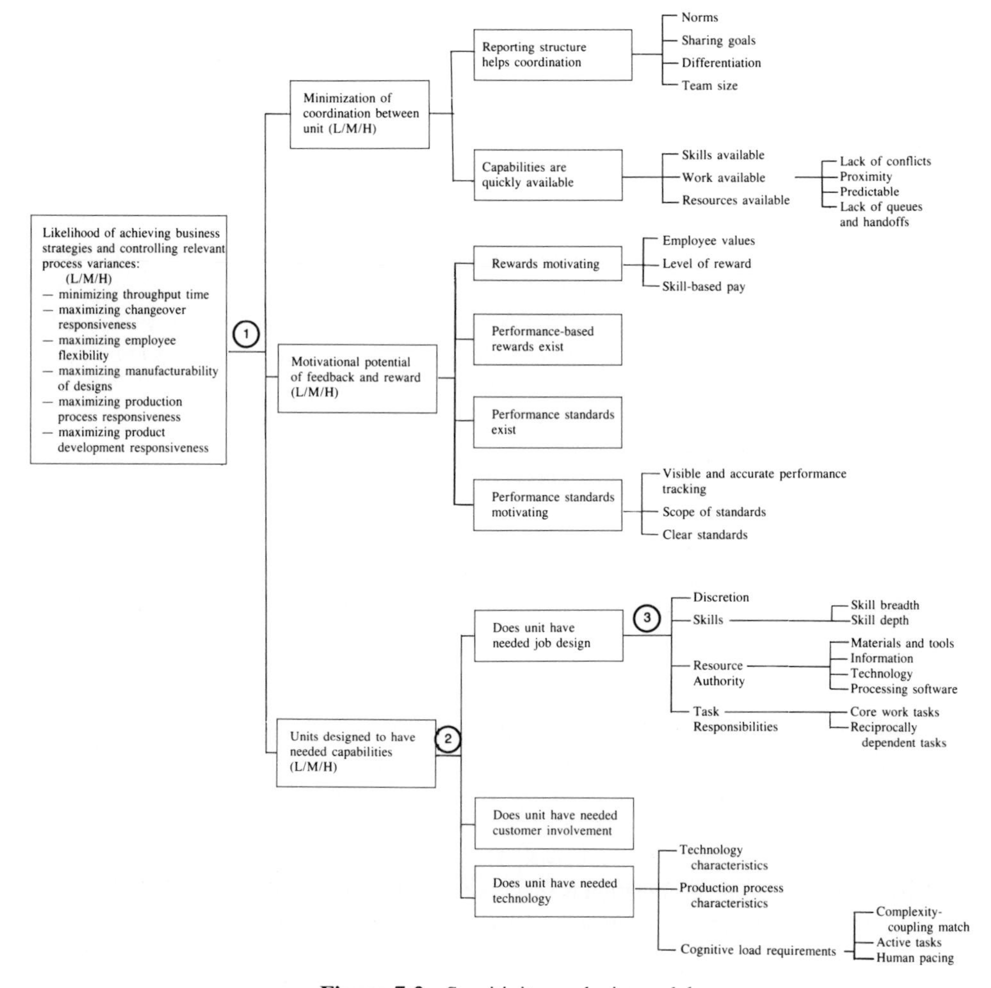

Figure 7.2 Sensitivity analysis model

- The degree to which aspects compensate each other has a lower threshold; for example, experts outside the work role cannot compensate for a work role in which people in that work role are so poorly skilled that they cannot make use of the experts.

These assumptions are clearly simplifying assumptions and thus, may not prove to be valid with further development. However, the software and knowledge representation for these assumptions are user-changeable so that as case experience accumulates, more sophisticated propositions about weighting and compensatory relationships can be devised. For the industry user, this model is intended to be an analysis of sensitivity. A user can input a known design to determine if deviations in the design from ideal relationships are so great as to harm organizational effectiveness. Or the user can experiment with alternative design features to determine which ones are most likely to have the greatest benefit for increasing organizational effectiveness, given the degree to which other design features have been constrained. For academics interested in the question of the relative importance and thresholds of the full spectrum of sociotechnical relationships when considered as a complete package, this model is intended to be a way of exploring all those relationships simultaneously and of experimenting with different weights and combination rules.

ACTION software

The knowledge base is large and thus a user interface that allows for easy navigation and understanding of the knowledge is essential. The easy navigation is managed by software that provides:

- immediate feedback to the user when ideal relationships are violated (by portraying the relationships as matrices and showing red squares when the user inputs information about his or her organization that violate the ideal relationships);
- summary reports that provide comprehensive lists of ideally needed system features and gap analyses in which the ideal features are compared with the features provided in the organization or proposed design;
- four different modes for using the system to correspond to different types of users (i.e. users wanting to access the knowledge base at a summary versus detail level; and users wanting to use the knowledge base to evaluate a fully-scoped design or existing organization versus using the knowledge base to fill in design parameters not yet defined).

The software is intended to help sociotechnical designers, be they managers, engineers or shop-floor workers, to analyse their current or proposed operations to assess the adequacy with which it integrates among technology, organization, and people issues, as well as to help identify new design choices. ACTION users may be analysts in the organization such as industrial engineers, manufacturing engineers, or organizational change analysts. Alternatively, ACTION users may be managers, such as production managers, operations managers, or plant managers. Finally, ACTION could be used by shop-floor people empowered to perform continuous improvement. A figure describing what ACTION, the software, does relative to the role of people in the sociotechnical design process is presented in Figure 7.3. The software has been operational on Sun, DECs and HP work stations since February, 1993.

Case example of industry use of ACTION

Case runs with the ACTION system have been conducted by the industry partners to validate the knowledge base and test the applicability of the knowledge and process in real world settings. Runs have been made on a wide variety of manufacturing contexts including: electronics fabrication and assembly, automotive fabrication and assembly, and metal fabrication.

Case background

The case described below was conducted in a metal fabrication facility which produces parts for the defence industry.

The industry as a whole is experiencing significant changes including: substantial decreases in the defence budget, continued stretch out of defence programmes, movement of contracts from higher to lower volume, greater competition from commercial shops,

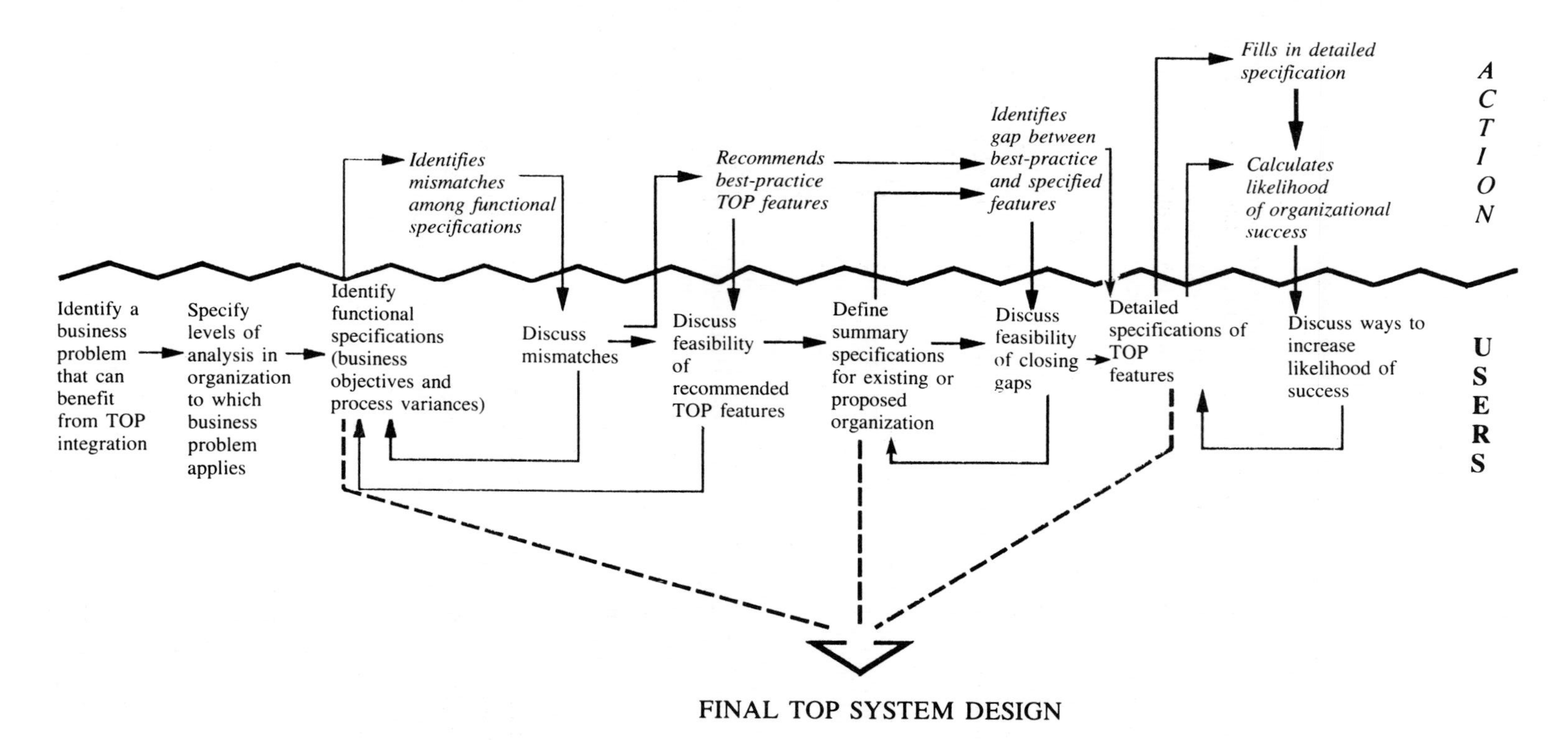

Figure 7.3 How ACTION facilitates people designing TOP* systems

* Technology, Organization & People

and increased environmental standards and regulations. Despite winning Baldrige awards for quality and accolades as a best practice organization, the drastic reduction in defence expenditures has driven the company to down-size their defence business, bringing it in line with expected levels of defence spending. At the same time, this facility struggling to become more competitive with other commercial fabrication shops that have lower overheads. These drivers are forcing the facility to make two major organizational changes.

The first is a reshuffling and regrouping of work load and equipment into Business Units (BUs). A BU is a process-based cell in which different equipment with different processing capabilities across similar products are grouped together. In a BU, the manager is responsible for overall management of operations which includes centralized customer contact, personnel management, scheduling, and process improvements. The cell includes about 60 machinists across three shifts with one technician/scheduler and one production engineer.

The second change is the grouping of machinists into self-directed work teams, in which machinists are cross-trained across different products and processes within the area, production engineers and technologies are allocated to each team, and the role of supervisor is one which is transitioning to that of a team facilitator. The 60 machinists are grouped into a total of six teams, three on the first shift, two on the second shift, and one on the third shift. Both changes are driving cycle time and cost reduction efforts.

To implement these two changes, facility managers have been relying on traditional methods. These traditional methods include addressing organizational and people issues only on an as-needed basis (i.e., trial and error) rather than planning ahead before problems arise. In addition, training has been focused solely on communication and working in teams rather than more broadly based skill training. Little emphasis has been placed on analysing and brainstorming needed changes to such larger organizational structures as reporting hierarchies and compensation systems. As a result, so far, changes in the Business Unit have been narrow in scope, with few significant changes in job positions, reporting relationships, and process technology. Facility managers have sensed that such a narrow scope and trial and error method for implementing organizational, technology, and people change yields limited success. They have learned from past efforts that silver bullets such as predetermined team structures, statistical process control, and cycle time reduction efforts have met with limited success because they involve organizational-wide change, and such change is an extremely complex process with an overwhelming number of factors to consider. Understanding this, facility managers have supported the development of a tool such as ACTION. The case described here was the beginning of a proof of concept. The focus in this assessment was to use ACTION on one BU within the facility and at only a summary level. The areas of improvement pointed out by the modelling of the one unit can then be utilized by the remaining business units at the facility.

The unit modelled in this case is responsible for rapid reaction machining of low volume, medium tolerance, metal parts. The manufacturing process includes numerical control and manual machining, drill, deburr, and inspection. The unit also has responsibility for developing their own process methods, numerical control programmes, and scheduling.

The case has been developed by the site industrial engineer working with the Business Unit Manager. The industrial engineer was responsible for operating the ACTION software and consulting on the ACTION theory. The Business Unit Manager, or process owner, provided system inputs and worked interactively with the software.

Case detail

The case study process began with gathering information about the business unit from a wide range of knowledge owners. The data was then entered into ACTION, running the system interactively with the Business Unit Manager. Recommendations were recorded and validated against manufacturing and company realities. Changes in inputs were then implemented and learning about the ACTION process recorded. This particular case was run at a summary level of the Business Unit's operations. The goals of the run were to provide feedback to the Business Unit Manager on how to handle and best implement the changes of a Business Unit structure and teaming, and to test the usability and theory of the ACTION system.

This ACTION case run began with the identification of the business unit's business objectives. In this example, the business unit manager selected his business unit's objectives to be:

- minimizing throughput time (interpreted as reducing cycle time);
- maximizing quality;
- maximizing employee flexibility for teams of generalists;
- maximizing manufacturability of designs;
- maximizing changeover responsiveness (interpreted as reducing set-up time).

These objectives were selected as those that most directly support the entire facility's competitive strength. Upon selecting objectives, ACTION checked the set of objectives to make sure they were in alignment with each other. One common way objectives are misaligned is that different objectives necessary to support each other are not in place. For example, throughput time can not be as readily reduced if quality and manufacturability of designs are not emphasized as well. Another common way that objectives are misaligned is that they refer to different units of analysis. That is, if the unit is focusing on minimizing throughput time for a portion of the production process, then the ACTION theory would suggest that quality should also be focused on a portion of the production process, and not individual work stations. In this example, no misalignments between business objectives was noted.

The next step in the summary use of ACTION was to select process variances and align them to the objectives. In this case, the following variances of incoming quality, output quality, scheduling changes, equipment accuracy, part design quality, and machine/cell assignment were identified as problems requiring considerable time and energy in rework, extra coordination, and equipment downtime. The ACTION system juxtaposes identified variances against business objectives to clarify the negative impacts of the variances on the business unit's ability to achieve its objectives. When ACTION graphically displayed the effects of variances on objectives, the Business Unit Manager could more clearly understand why he was having difficulty meeting his objectives of reduced throughput time and improved quality.

The third step in using ACTION is for the user to describe the BU in terms of such features as skills provided, information resources provided, customer involvement level, employee values, discretion, technology and process characteristics, and performance measurement characteristics. In this case, the BU included skilled machinists, a cell manufacturing engineer, and a cell production technician. The skills represented by these three jobs were substantial, including designing processes, operations, and scheduling. There was little cross training among the different positions. Most information resources except for customer needs and product costing are provided to all of the jobs through computerized systems in the area. There is little direct involvement of customers in the BU since the manufacturing planning department provides direction on long-term scheduling and manufacturing improvements. Current employee values support continuous improvement and a learning environment. Discretion is provided to employees over the design and sequencing of the manufacturing process. The production process is in large part manual and can be described as reliable, with breakdown alternatives and human override. The performance management system is one in which the cell-team facilitator evaluates each machinist based on his or her individual performance. Pay is based broadly on quality of performance and on more general areas such as team work, attendance, and safety. All of this information about the BU was entered into the system over a 3-week period with a total amount of effort of about a day. Once these descriptors were inputted into the system, they were matched to the features needed in the BU if the BU is to achieve its intended objectives. The results of this comparison of needed to actual is conveyed on the screen as a gap analysis. It is this analysis that helps determine what additional features the BU must consider implementing to achieve its business objectives.

Summary of gap analysis findings

There were four key gaps identified for this BU. First, several different gaps pointed to the need for the business unit to have greater direct customer involvement. To achieve its objectives of minimizing throughput time and maximizing quality, the ACTION theory suggested that the BU will benefit from having resident knowledge about the customer's work processes, being provided with continuously updated information about the customer's needs, and involving customers in such business issues as long-term scheduling and evaluating the BU's performance. The required knowledge, information and customer involvement were not provided in the BU.

A second key gap finding was focused on the unit's performance measurement and reward systems. In order for the selected objectives to be achieved, ACTION theory states that BU employees need to know where they stand on the path to achieving the unit's objectives and feel that they will be appropriately rewarded for that achievement. In this example, because the objectives were broader than those achievable by an individual worker, the ACTION run indicated that the business unit needs people for accomplishing work beyond that achievable by individuals working alone. Thus, performance measurements and rewards need to include team and unit-based compensation. In addition, the standards for performance need to be clearly communicated so that the entire unit can understand the connection between how they as individuals are being measured and the achievement of the unit's objectives. These features were not currently provided in the BU.

A third gap area identified concerned information. The ACTION theory suggests that to reduce throughput times, updated information on product and process flow and the capabilities of each employee to meet these flow requests is needed. This information was not currently available to BU workers. In addition, ACTION indicated that information needed to be distributed to all jobs in the BU. Currently, information was only selectively provided to certain jobs.

The fourth gap highlighted a need for the BU to be involved in redesigning and developing new parts. Traditionally, the facility as a whole has been isolated from the design community with new designs being 'thrown over the wall' to manufacturing. As manufacturability of designs was an objective of the BU, the features provided in the BU generated many red squares (i.e. were insufficient to meet the objective as specified in ACTION theory). For example, relationships with customers were insufficient to create an understanding of customers' needs to make part design decisions that simultaneously consider customer and manufacturing needs.

Recommendations to BU manager

Based on the outputs from the ACTION run, the Industrial Engineer suggested several recommended actions to the Business Unit Manager. To improve the first gap of customer involvement, three recommendations were developed. First, operator visits to the customer's work site were suggested. These visits would provide a better understanding of how the customer is using the business unit's product, allow the machinists to hear firsthand about what is really important, and allow them to witness the impact of their work on the bigger picture. One such visit has been conducted in the past and proved to be an outstanding learning opportunity for the operators as well as the customers. Continuing these visits on a regular basis is recommended. A second proposal to improve customer involvement was to feed the details of the customer survey results down to the business units. Typically, these results are reviewed by the planning organization and not shared with the production areas directly responsible for the product. Finally, it was also suggested that the business unit develop their own customer survey to gather direct feedback about their operations and product.

To improve the performance measurement system, two recommended actions were suggested. The first was to implement a team-based review process. The cell teams in this unit are now advanced enough to take advantage of this method of review and efforts are underway on a broader level to implement a team-based reward system across the division. The second recommendation was that the unit should focus on the key business objectives of the organization, posting the specific numeric standards so that all can see where they are and where they need to go. Most importantly, they need to be judged on their achievement of these goals. Efforts have already begun towards tracking progress on specific business objectives. Throughput time and scrap are now tracked at the team level providing information to direct process improvements.

Three recommendations were made to close the gap about the need for information. The first was to train the teams in the techniques of process mapping. This would provide each team with the tools to map their manufacturing process and thereby better understand the product and process flow. Second, a personal computer based program was created to track individual operators' capabilities and what additional training they wanted. With this knowledge, the teams and the Business Unit Manager would have the information they needed to move people around to fulfill specific manufacturing requirements. The third recommended change was to distribute additional computers around the shop to provide easier access to existing information sources. This change has also been implemented.

Finally, the fourth gap indicated a need to create tighter links to the design community. To improve this linkage, the recommendation was made that the BU should coordinate directly with project management and the design engineers on issues such as manufacturability and production cost estimating, instead of going through a series of middlemen (called manufacturing engineers). This link needs continued strengthening in the future to allow continuous communication rather than just in a few special cases.

In summary, the ACTION run provided enough detailed discussion to generate nine specific recommended actions within 4–5 hours of work interpreting the results. Of the nine recommendations, five have already been implemented and four are under discussion with a more specific implementation plan under development.

Case conclusions

Several conclusions can be drawn concerning the ACTION process. First, we found the interactive nature of the system to be very valuable. The impact of the technology, organization, and people trade-offs were more clearly understood by management when they could see immediate response to their changes in inputs on the screen. Second, the Business Manager noted that the gaps identified were more believable coming from an engineer using a well-developed knowledge base of academic research and experiences of best-practice cases than from the engineer alone. Third, we concluded that at this time, the ACTION system needs an 'expert' user (e.g. the engineer) in order to facilitate its use and translate its output. Fourth, we found that the potential benefit of ACTION will derive in part from how inputs are gathered. In this case, inputs were provided only by the Industrial Engineer working with the Business Unit Manager. This might have resulted in a less complete or less accurate picture of the current operations as well as failed to leverage the benefits of having a broader base of individuals discussing their different perspectives on the inputs. In future applications, we recommend including a broader community such as operators, cell engineers, and customers to stimulate those frank and open discussions that often create the greatest insight to the production area. Fifth, it was noted that only one of four modes for using ACTION was addressed in this case. With more detail inputted into the system, the expectation was that even more information would have been provided.

Overall, the recommendations and insights provided by ACTION process proved valuable. However, they could be considered common sense (note that summary-level use of ACTION does not include the sensitivity analysis). What is important is the interactive ability to balance **all** of the technology, organization, and people issues against each other. During the typical change process, it would be difficult to consider all of these factors to come up with a complete and comprehensive solution. In this example, the Business Unit Manager was not particularly surprised by the results. The ACTION recommendations seemed to make sense. He did however, realize that many of the important issues would have slipped through the cracks without the help of a defined theory, methodology, and automated system. Without ACTION, he would have counted on the 'silver bullet' of teaming and business unit alignment alone to solve everything. Now the bigger picture is understood and corrective action can begin.

Conclusions

Accumulated evidence indicates that the implementation of computer-aided technology has not achieved as much success as originally anticipated (Majchrzak, 1992). Failures or problems with new technology can occur for a variety of reasons: technical barriers, inadequate skills, inadequate resources, etc. However, accumulated evidence indicates that the major source of problem is the inadequate planning for integrating the technology, people, and organizational change (Office of Technology Assessment, 1984;

Criswell, 1988; Dertouzos *et al.*, 1989; Grayson, 1990; Ozan and Smith, 1990).

One major reason why integrative planning is so difficult is that existing tools are inadequate (Majchrzak and Gasser, 1991). Simulation packages focus exclusively on technical-design choices or very high-level human-resource choices (such as workforce size or general workforce composition). Business re-engineering tools provide mechanisms for a user to build his or her own model of the organization; however, there is little guidance for using the model to create organizational and job designs. Sociotechnical design tools are heavily resource-dependent and time-lagged, which makes the process of designing solutions slow.

Ideally, integrative planning will be easier when a tool exists that contains a knowledge base of how technical-design decisions impact organizational and people issues, and vice versa. In addition, such an ideal tool would also be computer based to allow for the simultaneous generation of alternative technology, people and organizational design decisions. Such a tool, under development for the last several years, is called ACTION.

Acknowledgements

This paper was prepared as a result of a programme sponsored by the National Center for Manufacturing Sciences. This is a joint R&D programme among industry and academia. The authors would like to thank Les Gasser for his contribution to this work. Requests for reprints should be addressed to Ann Majchrzak, University of Southern California, ISSM-COD Lab, 927 West 3th Place, Room 116, Los Angeles, CA 90089-0021, USA.

References

Carter, D.E. and Baker, B.S. (1992) *C.E., Concurrent Engineering: The Product Development Environment for the 1990s*, Reading: Addison-Wesley.

Champion, M.A. (1983) Personnel selection for physically demanding jobs: review and recommendation, *Personnel Psychology*, **36**, 527–550.

Clark, K.B. and Fujimoto, T. (1991) *Product Development Performance: Strategy, Organization, and Management in the World Auto Industry*, Boston: HBS Press.

Criswell, H. (1988) 'Human System: The People and Politics of CIM', paper presented at the AUTOFACT Conference, Chicago, May 28–30.

Dertouzos, M.L., Lester, R.K., Solow, R.M. and the MIT Commission on Industrial Productivity (1989) *Made in America: Regaining the Productive Edge*, Cambridge, MA: MIT Press.

Duimering, P.R., Safayeni, F. and Purdy, L. (1993) Integrated manufacturing: Redesign the organization before implementing flexible technology, *Sloan Management Review*, **34** (4), 47–56.

Eason, D. (1988) *Information Technology and Organizational Change*, London: Taylor & Francis.

Gael, S. (1984) *Job Analysis*, San Francisco: Jossey-Bass.

Gagne, R.M. (1985) *The Conditions of Learning and Theory of Instruction*, New York: Holt.

Grayson, C. (1990) 'Strategic Leadership', paper presented at the Conference on Technology and Future of Work, Stanford.

Hackman, J.R. (Ed.) (1990) *Groups that Work (and Those that Don't): Creating Conditions for Effective Teamwork*, San Francisco: Jossey-Bass.

Hammer, M. (1990) Reengineering work: Don't automate, obliterate, *Harvard Business Review*, **68** (4), 104–112.

Hammer, M. and Champy, J. (1993) *Reengineering the Corporation: A Manifesto for a Business Revolution*, New York: Harper Business.

Hayes, R.H., Wheelwright, S.C. and Clark, K.B. (1988) *Dynamic Manufacturing: Creating the Learning Organization*, New York/London: The Free Press/Collier Macmillan.

Lawler, E.E. (1988) *High-Involvement Management*, San Francisco: Jossey-Bass.

Liker, J., Majchrzak, A. and Choi, T. (1993) Social impacts of programmable manufacturing technology, *Journal of Engineering and Technology Management*, **10** (3), 229–264.

Majchrzak, A. (1988) *The Human Side of Manufacturing Automation: Managerial and Human Resource Strategies for Making Automation Succeed*, San Francisco: Jossey-Bass.

Majchrzak, A. (1992) Management of Technological and Organizational Change, in: Salvendy, G. (Ed.), *Handbook of Industrial Engineering* (second edition), New York: Wiley, pp. 767–797.

Majchrzak, A. and Gasser, L. (1991) On using AI to integrate the design of organizational process change in U.S. manufacturing, *AI and Society*, **5** (4), 321–338.

Manufacturing Studies Board (Committee of Effective Implementation of Advanced Manufacturing Technology, National Research Council, National Academy of Sciences) (1986) *Human Resource Practices for Implementing Advanced Manufacturing Technology*, Washington: National Academy Press.

McCormick, E.J. (1976) Job and task analysis, in: Dunette, M.D. (Ed.), *Handbook of Organizational and Industrial Paychology*, Chicago: Rand McNally.

Nagel, R. and Dove, R. (1991) *21st Century Manufacturing Enterprise Strategy*, Bethlehem: Iacocca Institute.

Office of Technology Assessment (1984) *Computerized Manufacturing Automation; Employment, Education and the Workplace*, Washington: US Government Printing Office.

Ozan, T.R. and Smith, W.A. (1990) *American Competitive Study: Characteristics of Success*, Chicago: Ernst and Young.

Pava, C. (1983) *Managing New Office Technology: An Organizational Strategy*, New York: Free Press.

Porter, M.E. (1985) *Competitive Advantage*, London: Free Press.

Senge, P.M. (1990) *The Fifth Discipline: The Art and Practice of the Learning Organization*, London: Century Business.

Smith, P.G. and Reinertsen, D.F. (1991) *Developing Products in Half the Time*, New York: van Nostrand Reinhold.

Stalk, G., Evans, P. and Shulman, L.E. (1992) Competing on capabilities: the new rules of corporate strategy, *Harvard Business Review*, **70** (2), 57–69.

Stewart, T.A. (1993) Reengineering: The hot new managing tool, *Fortune*, **128** (4), 41–48.

Taylor, J.C. and Felten, D.F. (1993) *Performance by Design: Sociotechnical Systems in North America*, Englewood Cliffs: Prentice Hall.

Thompson, J.D. (1962) *Organizations in Action*, New York: McGraw-Hill.

Trist, E.L. and Murray, H. (Eds) (1993) *The Social Engagement of Social Science. A Tavistock Anthology. Volume II: The Socio-Technical Perspective*, Philadelphia: University of Pennsylvania Press.

Winby, S.S. (1993) The high performance workplace: Managing people and technology in the 21st century, *Operations Management Review*, **9** (2), 37–41.

Woodward, J. (1982) *Industrial Organization*, London: Oxford, Oxford University Press.

8

Artificial Intelligence and Knowledge-based Systems: A New Challenge for the Human-centred Perspective?

John Kirby

Introduction

In the introduction to this book it was observed that, despite many attempts to develop human-oriented symbiotic approaches to systems design, the traditional technology-centred approach and culture still dominates. A central characteristic of this traditional approach is the incorporation of as many functions as possible into the system leaving the operator to carry out activities which are too difficult or too costly to automate. Artificial Intelligence technology promises to extend the range of human activities which may be incorporated into the systems by seeking to automate 'intellectual' activity. This seemingly poses new problems and challenges to researchers intent on developing alternatives to the dominant approach of system designers and developers.

A human-centred approach is advocated in this chapter and is based on the belief that systems which incorporate the need for skilled and knowledgeable human action are more effective as well as having less detrimental effects on the quality of working life of their operators (Rosenbrock, 1977). This is the opposite of the traditional technology-centred approach which is motivated by the belief that closely supervised and controlled semiskilled or unskilled labour is more efficient than skilled labour (Landes, 1969). The result is deskilling and the removal of control of the work process from the operator, who often becomes an appendage of the machine. For Rosenbrock (1989), the goal of the human-centred approach is the development of technical systems in which the operator is not subordinate to the machine. This means the design and development of technical systems which seek to support the role and enhance the skill and knowledge of their users or operators. The development of such systems requires not only the reorganization of work but also a corresponding fundamental transformation of the approach to technical design and development which allows for the consideration of a broad range human issues. In common with other symbiotic approaches, the human-centred approach has sought to address these human issues by the creation of design teams consisting of social scientists as well as technologists.

Artificial Intelligence (AI) as an academic discipline is concerned with the development of computer systems to emulate human abilities in natural language understanding, visual perception and problem solving. A significant part of AI research has centred on the development of ways of understanding, representing and incorporating human knowledge into computer systems to create knowledge-based systems (KBS). These developments in AI/KBS have opened up the possibility of building computer systems which may be capable of carrying out complex tasks involving in-depth understanding, skill and judgement. In so doing AI/KBS increases the potential for further automation of production and office work and the extension of automation to areas of work traditionally carried out by professional personnel, such as doctors and engineers.

The most well-known and widely used result of academic AI/KBS work is expert systems which is the application of research in the area of problem solving and reasoning. Developed by Shortliffe in the early 1970s, the medical diagnosis system MYCIN was the first system recognizable as what would be understood today to be an 'expert system' (Shortliffe, 1976). Perhaps more important than being the first has been the fact that MYCIN established the basic paradigm for expert systems. This paradigm would appear to be characterized by a number of key objectives:

- To replicate areas of professional intellectual activity central to the primary role of the professional group.
- To achieve expert level performance in the chosen areas of intellectual activity.
- To use an automated problem solving approach where the user enters data about which the system reasons and produces a solution.
- To represent heuristic human expert knowledge in the form of IF . . . THEN . . . rules.

The expert systems paradigm established by MYCIN as outlined above has had a wide influence on thinking about expert systems especially outside the AI community.

From a human-centred perspective there are a number of problems with the expert systems paradigm. First, by attempting to achieve expert level performance on tasks central to the role of a professional group, expert systems hold out the possibility of replacing key aspects of their professional role. Second, the role of the user of such a system would be reduced to that of entering data requested by the automated reasoning system. This would effectively make the user subordinate to the system and represent a deskilling of the work process. Third, the motivation, or at least the justification, for the development of expert systems is typically the alleged inadequate performance of the professional group. In relation to MYCIN this was based on the observation that doctors prescribed more antibacterial drugs than would be expected from the number of people suffering from bacterial infections. It was alleged that the reason for this 'failure' was that doctors adopted an irrational and unmethodical diagnostic and prescribing procedures. However, this is not the only or even most likely explanation of this phenomenon as there are many complex reasons why doctors prescribe antibacterial drugs for patients who turn out not to be suffering from bacterial infections. In all three respects the expert systems paradigm appears to be very similar to the traditional technology-centred approach to the introduction of technology.

Expert systems are often described as 'decision-support systems', 'knowledge-based decision-support systems' or 'advisory systems'. However, it is unfortunate that the description 'decision-support system' is used as a synonym for expert system or other artificially intelligent system. In this chapter a broader view of decision-support systems is proposed which places the emphasis on assisting the user to arrive at decisions. Computer tools for use by professional and managerial staff provide a limited amount of this form of decision support. Spreadsheet and project management packages, for example, assist decision making by providing better ways to structure and present information which allow the user to explore different scenarios and options. Here decision support is provided in a way which supports the role and enhances the skills and knowledge of users. Is it possible to use AI/KBS technology to provide this kind of decision support rather than attempting to usurp the role of the user? This is the challenge which has been addressed in different ways by both of the projects described in this chapter. As well as having implications for the development of a human centred framework for the use and development of AI/KBS technology, this work raises a number of general issues for symbiotic approaches.

This chapter consists of four sections. The next section describes an application of the basic human-centred philosophy to the design of knowledge-based systems in computer-aided design for control engineers. This is followed by a presentation and discussion of the PEN&PAD (Practitioners Entering Notes & Practitioners Accessing Data) project which employed a user-centred design process in the development of the use of AI/KBS techniques in clinical patient record systems. After this there is a discussion of some implications of the work described for the human-centred approach in terms of the purpose of systems, the design process and knowledge representation. Finally, some conclusions are presented.

Knowledge-based systems in control systems design

The original aim of the work described in this section was the development of an expert system for control engineers using a computer-aided design package. The established framework for using expert systems in this context was oriented to automating the design process. In contrast to this established framework, the primary objective of the work described here was to find a role for an expert advisory system which would support the role and enhance the skills and knowledge of the control engineer. However, a further investigation of problem area led to the conclusion that an alternative framework for knowledge-based advice giving was need. It further became clear that meeting the requirements of this alternative framework would be difficult, if not impossible, to realise using expert systems technology and that a different technical solution would have to be used.

In this section the established framework for the use of expert systems in computer-aided control systems design is described along with some of its attendant problems. This is followed by the proposal of an alternative general framework for knowledge-based advice giving. Finally, there is a description of the use of this alternative framework in the development of an advisory system for use by control engineers using a computer-aided design package. But first it is necessary to set the scene by briefly describing the role of control engineers and how they use of computer-aided design packages.

Control engineering design

The main role of a control engineer is to design a device known as a controller or control system which acts to modify the behaviour of an existing piece of plant or machinery. Controllers may be electronic, mechanical, pneumatic or hydraulic in operation and they are often associated with a digital computer. Plant or machinery for which controllers are developed include aircraft, spacecraft, ships, chemical and nuclear installations, and mining machinery. One example of a control system is the autopilot of an aircraft. It can

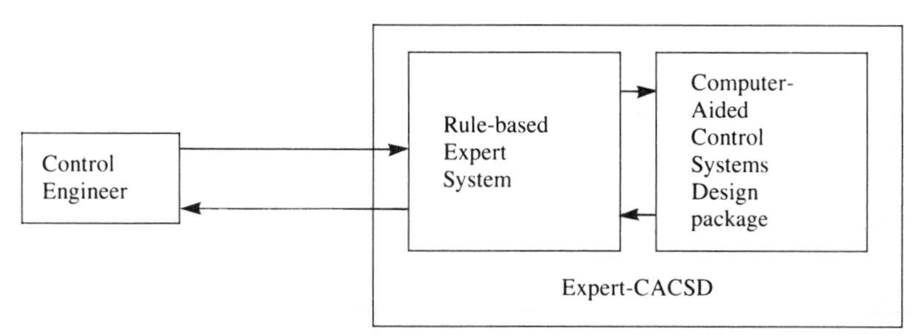

Figure 8.1 Expert-CACSD

be seen from this list of applications that control engineering is a highly responsible profession involving wide-ranging implications for public safety.

Control systems design is highly mathematical and the resulting design of a controller is usually expressed as a mathematical equation. Computer-aided control system design (CACSD) packages have been developed to assist control engineers in the design process. Because of the mathematical basis of control systems design, CACSD packages provide the engineer with the numerical algorithms with which to manipulate complex mathematical equations. The design process usually involves the generation and manipulation of graphs based on these equations. These graphs constitute intermediate stages in the design process and are very important in helping to give the engineer a feel for the behaviour of the system to be controlled. The final output from a CACSD package is a mathematical equation for the desired controller.

Automating control engineering design

A decade has passed since it was first suggested that knowledge-based systems could have a role in CACSD. This original proposal by Taylor and Frederick established a framework for the use of MYCIN-like rule-based expert systems in CACSD (Expert-CACSD) which has been followed by most workers in the field (Taylor and Frederick, 1984). The basic idea is that of an overall expert system-controlled design environment in which the expert system mediates the interaction between the user and the CACSD package, as shown in Figure 8.1. In the typical operation of an Expert-CACSD system, the user enters a mathematical equation of the system to be controlled along with data about its desired behaviour. The expert system interacts directly with the CACSD package receiving and responding to intermediate results until the mathematical equation of the required controller is produced. At this point the controller design is presented to the user who can either accept it or modify the data about desired behaviour and repeat the operation.

From this description it is clear that Expert-CACSD is essentially an extension of the expert systems paradigm: controller design is part of the primary role of the control engineer; the aim is expert level performance; an automated problem solving approach is used in conjunction with a CACSD package; and knowledge is represented in the form of rules. As a consequence, the design process is effectively deskilled with the role of the user being reduced to that of entering data. The motivation for the development of Expert-CACSD also has echoes of the expert system paradigm. The primary justification for the development of this approach has been the assertion of the existence of a class of less-than-expert-users of CACSD. One sense in which users might be less than expert is in relation to their familiarity with particular CACSD packages which are often idiosyncratic and difficult to learn and use. However, little of the work reported on Expert-CACSD has been concerned with these usability issues. The main meaning of the notion of less-than-expert-user has been that there is a class of control engineers who are in some way less than expert at their job.

Responses to Expert-CACSD from control engineers

There has been a strong reaction from control engineers against automating control-systems design. Four years after the original proposal, Taylor (1988) had to concede that there were a number of problems with the Expert-CACSD approach:

- User involvement in the design process: Because control of the design process had been taken over by the expert system, the designer does not receive or have to think about the intermediate results. Thus the engineer would develop no empathy with the behaviour of the system or its proposed controller.
- Responsibility: One very strong feeling was that many of the automated functions were too important to be left to the expert system to make judgements. Too important both in the narrow terms of the design process and in the broader social implications of producing controllers for use in aircraft, nuclear installation etc., where the design decisions had been made by an expert system.
- Deskilling: If the typical user of Expert-CACSD was a less-than-expert control engineer then this would lead to deskilling. Acknowledging and responding to this criticism Taylor states that: 'The first consideration in our work has been that the user of (Expert-CACSD) software would be a control engineer (no — we are not yet ready to propose that your job be eliminated!)' (Taylor, 1988: 11).

As a consequence of the responses of control engineers, Taylor proposed a much less ambitious role for expert systems in CACSD. Instead of an overall expert system-controlled CACSD environment, he proposed that only certain aspects of the design process were suitable for automation (Taylor and McKeehan, 1989). However, this new approach seems only to represent a change in the proposed scale of involvement of expert systems rather than any fundamental redefinition of their role (Kirby, 1991).

An alternative approach to knowledge-based advisory systems

The framework for Expert-CACSD had been derived from the general expert systems paradigm established by MYCIN. Consequently, in the search for an alternative, it was necessary to look for a different general framework for knowledge-based advisory systems such as the 'alternative paradigm' provided by Coombs and Alty (1984). They argue that MYCIN-like expert systems cannot be described as advisors or assistants because they are, in fact, problem solvers in their own right. According to Coombs and Alty, studies show that human experts do not usually seek advice in their own area of expertise. Rather, they are more likely to seek the advice of an expert in a related subject area. Furthermore, such advice is rarely in the form of a complete solution. Human experts are more often called upon to provide guidance so that advised persons are given sufficient understanding to solve their own problems.

Based on this understanding of human expert-to-expert advice giving, the alternative paradigm is based on two main requirements:

1. Adjacent Knowledge Domain: The system should provide advice on a subject adjacent but relevant to the expertise of the person seeking advice. This may also mean a more specialized area within the same general subject area.
2. Knowledge Communication: The role of the advisory system should be to communicate knowledge so that the users can solve their own problems.

Both of these fit in very well with the objective of designing a system to support the role and enhance the skills and knowledge of the user. It is worth emphasizing that the alternative paradigm is concerned with expert-level operation of the kind attempted by MYCIN and Expert-CACSD. This is different from the many expert-systems' applications which have been developed to automate limited, routine and possibly irksome aspects of users' work.

In addition, the aim of developing systems in which the operator is not subordinate to the system necessitates the inclusion of a further requirement:

3. User Control: A human-centred advice-giving paradigm must provide for the greatest possible user control of the system and, in particular, user control or ownership of any decision-making or problem-solving activity.

The inclusion of this requirement extends the alternative paradigm proposed by Coombs and Alty and provides the basis for a human-centred framework for the development of knowledge-based advisory.

An advisory system for control engineers

In the application of the approach outlined above, the first issue considered was that of an appropriate subject area for advice giving; an area which would be adjacent but highly relevant to the control engineer using a CACSD package. From firsthand experience of providing support to users of a major CACSD package, it had emerged that one area of concern was that of the reliability and accuracy of the numerical algorithms or methods used. In fact, this is a well-documented issue for all CACSD packages (Laub, 1985). While control engineers have some understanding in this area, they would not usually consider themselves to be experts. It was, therefore, decided to develop a prototype system to give advice about the accuracy and reliability of the numerical methods used in a particular CACSD package, Numerical Methods Advisory System (NUMAS).

Expert systems are automated reasoning programmes, and, hence, effectively remove from the user control of the problem-solving process. In itself, this would be a sufficient reason to seek an alternative technical approach. Might an alternative be found which involves adapting expert-systems technology? The main problem here is that expert systems cannot effectively communicate the knowledge they contain. This was clearly demonstrated by Clancey (1983) who attempted to use MYCIN's knowledge base for teaching purposes. The first problem was that learners gained little understanding from the knowledge encapsulated in the rules. For rules to be comprehensible to the learner, the conceptual knowledge from which they had been derived had to be explained. The second difficulty was that learners obtained from MYCIN little understanding how they would themselves solve problems. Clancey concludes that MYCIN's 'lack of similarity to human reasoning severely limits the usefulness of the system for teaching purposes' (Clancey, 1983: 235).

An alternative technical solution was needed which is capable of storing and communicating knowledge in a form easily accessible to the user and which places the user in control of the interaction and the problem-solving process. This led to a consideration of hypertext technology, which has been described by Conklin as 'a computer-based medium for thinking and communication' (Conklin, 1987: 17). In a hypertext system text or graphics are displayed on the screen, usually in a window. Within the text or graphics certain words or areas, known as 'buttons', are highlighted in some way — colour, font or cursor shape. The user can point to a button using a mouse and when the mouse button is clicked other textual or graphical material is displayed on the screen. This new material may also contain hypertext buttons which can be clicked on to access yet more material. The details of the screen displays in hypertext varies from system to system.

In the development of NUMAS, the notions of procedural and conceptual knowledge developed by the AI/KBS research were used to structure and organize the material or knowledge base in a hypertext system. Procedural knowledge in the form of a decision tree was used to create links between descriptions of steps required to investigate a potential problem. Links were created between concepts and terms found within these procedural descriptions to definitions and explanations. A semantic network approach was used to create links between the concepts and terms. An initial prototype of a NUMAS was developed in order to investigate the technical feasibility of such an approach. Similar hypertext approaches using decision trees and conceptual knowledge have been used in the development of advisory systems which are now in regular commercial usage (e.g. Brown, 1989; Churcher, 1990).

Knowledge representation for medical information systems

A novel approach to the use of AI/KBS in decision support has emerged in the PEN&PAD project. When the project began in 1988 its aim was to investigate the application of AI/KBS techniques to the design of user interfaces to computerized patient-record systems used by doctors (Rector *et al.*, 1988). The incorporation of human issues into the design process was accomplished by the adoption of an iterative and participatory user-centred design approach. This close involvement with users led to a change in the focus of the project from the design of user interfaces to the use of AI/KBS techniques for representing medical terminology in a patient-record system.

This section begins with a description of the user-centred approach to design developed in the course of the PEN&PAD project. This is followed by a description and discussion of current approaches to representing medical terminology in medical information systems. Then, an alternative knowledge based approach to the representation of medical terminology developed in the PEN&PAD project is presented. Finally, there is a description of how this approach provides decision support. The presentation in this section goes into some technical detail. However, this detail is needed to illustrate the relationship between the underlying AI/KBS technology and how decision support is provided to the user.

Supportive evaluation methodology

One of the central components of the PEN&PAD project has been the development of a user-centred approach to design which became known as supportive evaluation methodology (SEM) (Rector *et al.*, 1992). Essentially, SEM is an iterative approach to design involving both users and social scientists in the design cycles. It appears to have similarities with the STEPS approach described by Latniak (Chapter 5). In the PEN&PAD project there were two groups of users: the 'Inner Circle' of six doctors who were regularly involved in the design process; and the 'Outer Circle' of another 16 doctors who were involved only at key stages of the design process. The evaluation team consisted of social scientists who had experience in the evaluation of medical computing systems. The development team consisted of academic and research staff in the Department of Computer Science at the University of Manchester.

Each cycle began with Requirements and Design Workshops involving Inner Circle doctors, the design team and the evaluation team. The workshops were organized, run and 'owned' by the evaluation team who saw themselves as facilitating a dialogue between the doctors and the design team. The pattern of these workshops was very variable but each was usually devoted to a specific design issue. Sometimes this involved general discussions, but, on other occasions, the discussions were focused on such things as: paper-based exercises, mock-ups of some aspect of the proposed system or some aspect of a prototype seen at a previous workshop.

Following a series of such Requirements and Design Workshops there would be a period of intense activity by the design team to develop a prototype system based on the ideas which had emerged from the workshops. During this period there was very little involvement of either the user group or the evaluation team. The work of the design team was certainly based on the understanding gained in the workshops but it also included the creative contribution from the designers. One important example here was the decision made by the design team to use the particular kind of data entry form which became one of the most important and widely popular aspects of the system. The design team made this decision because it appeared to be the best approach to addressing issues which had emerged from the earlier workshops. It was not something that was directly suggested by either the user group or the evaluation team.

At the end of the design period, Formative Evaluation Workshops were held involving the Inner Circle. Here, again, the evaluation team ran and owned the workshops and facilitated feedback from the user group on the current prototype. The format of these workshops was much more predictable. They began with a general introduction and demonstration by a medical member of the design team. This was followed by a period of hands-on training. The evaluation proper often involved the use of role playing and realistic patient scenarios. The doctors, working in pairs, each took a turn at playing the patient, while the other 'played' the doctor with the aim of simulating a real-life consultation as far as this was possible in an essentially laboratory setting. Following this activity, the doctors filled in questionnaires designed by the evaluation team before joining a general group discussion which was led by the evaluation team.

On the basis of the issues arising at the Formative Evaluation Workshops, the design team made modifications to the current prototype for a further workshop. When a particular area of functionality was felt to be well developed, Formative Evaluation Workshops were held involving members of the Outer Circle of doctors. This was done to ensure that the ideas emerging from the Inner Circle were not the idiosyncrasies of the specific group of doctors involved in the Inner Circle.

Current approaches to representing medical information

In early Requirements and Design Workshops it became clear that what appeared to be user interface problems actually had another origin. In fact, the major problem was, and still is, the coded format in which patient data must be entered and stored in most medical information systems. Typical coding and terminology schemes contain medical concepts or terms which consist of a short phrase or 'rubric' and a unique alphanumeric code. An example of an individual term is:

Rubric: Severe fracture of the shaft of the femur
Code: 12345

To enter clinical data using such terms, the doctor first invokes a display of a list of rubrics and then makes a selection from the list.

There are two major problems with this approach to entering clinical data. The first is accessibility. This arises because current coding and terminology systems contain very little semantic structure to help the user. Data entry typically starts with the user typing a three or four character 'key-word'. A list of rubrics which, typically have to be retrieved by simple word search techniques, is then displayed. This list may be very long, may contain rubrics on a wide range of clinical topics and can often contain groups of rubrics which differ by only a few words. As a consequence it is often very difficult to find the appropriate rubric. The second major problem is expressiveness. Because the rubrics are compound and predetermined, they do not allow the doctor to easily record an accurate description of each patient and his or her treatment. The result is that what is recorded in coded form only approximately reflects what the doctor actually wants to record about the patient. Any further elaboration on a rubric has to be entered as free text.

It was evident from early Requirements and Design Workshops that providing an elegant or even 'intelligent' graphical user interface for entering traditionally coded data would not solve this problem. Part of any strategy to improve the interaction between doctors and their computerized patient-record systems would have to be the development of an alternative to the current approach to coding and terminology.

One approach which would certainly achieve flexibility might be to use only free text with doctors simply typing their notes directly into the computer without the use of coded information. From the users' point of view there are two main objections to a free text only approach. First, one of the main requirements of a clinical patient-record system is that it should contain classified and structured information This point will be discussed further in the section on compositional data entry and decision support. The second problem is that doctors consider extensive amounts of typing to be not only time consuming but also disruptive to the doctor-patient dialogue. Despite these objections, it has to be recognized that there will still be occasions when free text is necessary to supplement the structured data.

Knowledge representation and compositional data entry

Further discussion with the Inner Circle of doctors revealed that they would prefer to be able to build up complex descriptions of patients' problems from elementary terms. What these elementary terms might be can be illustrated by considering again the example of the compound expression: 'Severe fracture of the shaft of the femur'. This rubric contains a number of words representing simpler concepts: fracture, femur, shaft and severe. It

fracture *which* **has Location** femur
has Severity severe
has Anatomical Site shaft

Figure 8.2 Decomposition of a compound expression into elementary terms

(fracture *which* **has Location** femur)
has Severity [mild, moderate severe]
has Anatomical Site [distal end, shaft, neck, proximal end]
has Fracture Form [simple, spiral, greenstick, comminuted]
has Openness Character [open, closed]

Figure 8.3 Part of a GRAIL model for fracture of the femur

also contains a number of implicit relationships between these concepts: severity, location, and anatomical site. Using these elementary terms and relationships, the compositional representation of the rubric can be expressed in the form shown in Figure 8.2.

This way of representing relationships between concepts or terms is based on the Artificial Intelligence notion of a semantic network. The representation formalism originally developed in the PEN&PAD project has undergone further development in the GALEN project and is now known as GRAIL — GALEN Representation And Integration Language (Rector *et al.*, 1994). A software system has been developed to store medical knowledge represented or modelled in GRAIL. The result is the creation of a knowledge base of medical terminology which has been called a 'Terminology Base'.

Essentially, the Terminology Base contains two kinds of knowledge. First knowledge about the classification of terms, for example, that the femur is a type of long bone which is a type of bone, or that breathlessness is a type of respiratory symptom which is a kind of symptom. The Terminology Base also contains knowledge about how terms may be sensibly combined together so that, for example, a fracture can be located in a bone but not in a kidney or lung.

Using the Terminology Base it is possible to provide truly compositional data entry interfaces allowing doctors to build up complex descriptions of patients' problems and treatments from simpler terms. This can be illustrated by considering the example of the complex concept a fracture of the femur. A GRAIL model containing some of the things it might be medically sensible to say about a fracture of the femur are shown in Figure 8.3. From knowledge represented in this way a data entry form can be generated as shown in Figure 8.4. The data entry task of the doctor is to select the appropriate words using a point and click device, such as a mouse or pen. Some selections are shown with thicker button borders and these would translate into more natural English as: 'A severe, open, simple fracture of the shaft of the femur'. It is worth noting that to achieve the same degree of expressiveness using the compound approach would require some 300 separate rubrics or up to 15 screens of information.

Compositional data entry and decision support

One very important consideration in the development of this compositional approach to data entry has been the desire to provide the doctor with genuine decision support. In the

Fracture of the femur

Severity:	mild	moderate	severe	
Anatomical Site:	proximal	shaft	neck	distal
Fracture Form:	simple	spiral	greenstick	comminuted
Openness Character:	open	closed		

Figure 8.4 Part of a compositional data entry form

first instance this is achieved by displaying on the forms the range of comments which are medically sensible to make about any given clinical topic. In this way the doctor is unobtrusively prompted and reminded of the relevant clinical factors in relation to a given topic. There is no requirement that forms have to be completed; the doctor can enter as much or as little detail as he or she considers to be clinically necessary for the particular patient.

Another way in which the classified and structured representation of patient data provides the basis for decision support is in relation to data presentation, manipulation and analysis. By providing the means of recording accurate descriptions of patients' problems a more complete and accurate clinical history is available to the doctor when making decisions about the patient. It may also help the doctor to be able to manipulate and analyse the patient's clinical history. For example, it may be clinically important for the doctor to review all occasions where a patient has reported any kind of respiratory symptom. This can be extracted from the clinical history by reference to the Terminology Base which contains knowledge about which symptoms are classified as respiratory systems.

This compositional approach to data entry and data presentation has been subjected to extensive user evaluation by the Inner Circle of doctors, the Outer Circle of doctors, and by doctors previously unfamiliar with the work. In general, the response has been very positive, with most of the doctors expressing the view that this approach is a great improvement on the current methods of representing clinical patient information. As a result of this success, a 3-year collaboration began in 1994 with VAMP Health Limited, the leading UK supplier of General Practice computing systems. The aim of this collaboration is the incorporation of the main results of the PEN&PAD project into VAMP's second generation system known as VISION. VAMP believe that to retain their leading position in this market, VISION must provide the kind of decision support and advanced data manipulation facilities developed in the PEN&PAD system.

Discussion

This section seeks to highlight some of the important differences between the approaches of the NUMAS and PEN&PAD projects and the approach of the prevailing expert systems paradigm to the application of AI/KBS technology. Foremost amongst these is the general and fundamental issue of defining the purpose and role of the system. This is the starting point of the discussion. This is followed by a consideration of the design process, which is also a general issue for the development of information technology systems. Finally, there is a discussion of the more specific issue of the choice AI/KBS technology and how it is applied.

The purpose of the systems

Decisions about the purpose of a proposed system are usually taken before the beginning of the design process proper. Karlsson (Chapter 4) points to difficulties in design-oriented research in relation to who defines the problem that the system is addressing. In particular, problems are often defined by top management and researchers with a consequent low acceptance of the results of the research by lower managers and workers. In order to avoid this kind of problem the purpose of the system should be based firmly on an understanding of real user needs in their overall work context. To determine whether or not a proposed system addresses such needs, it is necessary to consider two crucial questions:

- What problem is it proposed that the system should solve?
- How will the system provide the required solution?

These questions are not only relevant to the application of AI/KBS technology, but are also more generally relevant to the development of information systems. Of course, by asking these questions it is always possible that the answer is that AI/KBS techniques are not useful or even that no form of computerization is appropriate and that some other type of solution is required.

In the expert systems paradigm the problem to be solved is the alleged inadequate performance of a professional group on tasks central to the role of that group. Heathfield and Wyatt (1993) report that most medical expert systems reported in the research literature focus on the problem of diagnosis, which is a task central to the primary role of the doctor. However, they present strong evidence to suggest that doctors themselves do not seek computer assistance with diagnosis. The continuing preoccupation of developers of medical expert systems with diagnosis arises because the problem to be addressed is defined by the academic or the PhD student carrying out the research. In defining the problem 'developers do not adequately investigate the work practices of end-users, thus failing to identify those aspects of decision-making where support is genuinely required' (Heathfield and Wyatt, 1993: 2). The solution provided by such expert systems usually seeks to automate diagnosis and in so doing seeks to replace the skills and expertise of doctors. Given that the aim is to provide assistance in an area which is not wanted by doctors and in a manner which undermines their role, it is not surprising that very few diagnostic expert systems are in regular clinical use.

A completely different approach was adopted in the development of NUMAS. The identification of the problem was based on firsthand experience of dealing with users of a large CACSD package. The subject area of the numerical methods is not central to the role of control engineers, who are not expected to possess expert knowledge in this area. At the same time, an understanding of how to investigate the accuracy and reliability of numerical methods is highly relevant to the engineer's primary role as a designer. The purpose of NUMAS is to support the engineer's decision making in the design process by helping him or her understand the numerical accuracy of the results of design calculations. In relation to how this decision support was to be given, one approach might have been to develop a MYCIN-like expert system which would 'diagnose' the numerical accuracy of a calculation. However, this approach was rejected because such a system would not have been able to meet the requirements of knowledge communication and user control.

The problem addressed by the PEN&PAD project was the difficulties that doctors experienced when entering data into their patient-record systems. This had been identified as a key obstacle to doctors using such systems in an independent evaluation commissioned by the UK Department of Health. The overall purpose of these systems is the entry, storage, retrieval, and manipulation of clinical patient records; something that the majority of general practitioners have demonstrated their desire for by purchasing and attempting to use such systems. Originally it was thought that this improvement would be accomplished by providing an intelligent user interface to existing systems. However, it turned out that the problems experienced at the user interface had much more deep-rooted origins. The purpose of the AI/KBS technology developed in the PEN&PAD project has been to support and improve clinical information systems by providing a structured language for recording and manipulating clinical data. The development of this structured language has laid the basis for improved data entry facilities, which also provides the doctor with unobtrusive decision support.

The intended users of the applications of AI/KBS technology described in this paper are professional personnel. These influential professional groups have the power to resist, and often to veto, developments they find unacceptable. For example, the NHS provide financial support to general practitioners for the purchase of computer systems with the intention that these systems are used for recording clinical patient data. Despite this only a minority of general practitioners in the UK use their computer systems for recording clinical patient data — they are mainly used for administrative functions. As a consequence of this kind of power, the NHS and commercial suppliers are prepared to finance research and development to improve the acceptability of systems as the history of the PEN&PAD project clearly illustrates. This power and influence is in sharp contrast to the situation often encountered by unskilled, semiskilled, and even skilled personnel involved in manufacturing or clerical work. Here, systems are often developed with little regard to how acceptable they are to their operators. Far from having a role in determining the purpose of systems, operators often have new systems imposed on them and have little choice about whether or not to operate them. Badham (Chapter 6) describes human-centred projects which are attempting to challenge this sort of approach and to develop systems which are more acceptable to such operators.

The design process

It has been argued by Heathfield and Wyatt (1993) that most medical expert-systems projects have been technology-led. They suggest that the typical development process begins with the choice of a software tool usually in the form of an expert-systems development environment. The software tool then dictates the choice of amenable problems and the techniques which can be employed to solve the problems. There is little user involvement in the development process except for the elicitation of the knowledge of an expert. However, this expert may not be a member of the end user group and, in any case, the purpose of knowledge elicitation is the encoding of knowledge, not the elaboration of user requirements. Evaluation of the resulting system is focused on the accuracy of the diagnoses provided by the system with 'its impact on users and their clinical problems often ignored' (Heathfield and Wyatt, 1993: 3). There is rarely any involvement of social scientists in any stage of the process.

The approach of the NUMAS project was to attempt to apply a general 'human-centred philosophy' to a particular problem area. Although there was no formal involvement of social scientists in the development of NUMAS, there was an attempt to assimilate some of the lessons drawn by social scientists from earlier work (Corbett, 1985). On this basis three fundamental requirements were proposed as a framework for knowledge-based advice-giving systems: adjacency, knowledge communication, and user control. Here, the development process was requirements led, not technology led. Indeed, in the light of the requirements elaborated, the original intention to use expert-systems technology was abandoned in favour of a hypertext approach. The main weakness of the development process was the lack of sufficient user involvement and by the time this weakness had been identified it was too late in the project to remedy it.

The PEN&PAD project employed the supportive evaluation methodology (SEM) which involves both users and social scientists as well as system developers in the design and development process. SEM is a highly iterative and relatively informal approach in which the role of social scientists has been to facilitate a dialogue between representatives of the broad user group and the design team. This began at the earliest stages of the design process with the Requirements and Design Workshops and continued in the Formative Evaluation Workshops when prototypes had been developed. The aim was to keep the design team firmly focused on the tasks of users and their corresponding information requirements. It is important to remember, however, that in SEM only the design team are actually directly involved in making design decisions and carrying out design work.

In earlier research and development projects in manufacturing the emphasis was on creating design teams of technologists and social scientist, often with only minimal user involvement (Clegg and Symon, 1989). The aim of these projects has been the 'social shaping of technology' through the direct application of insights of social scientists in the technical design process. However, the relationship between the technologists and social scientists in such projects often has not been ideal. Rosenbrock (1985) perceived a particular problem of social scientists and technologists communicating effectively in the early tentative and conceptual stages of the design process. Other problems of social scientists working with technologists on advanced manufacturing systems have also been reported. Clegg and Symon (1989) reviewed the involvement of social scientists in ESPRIT projects on the development of human-centred CIM. They concluded that, despite their aim of direct involvement in technological design, the primary activities of social scientists had been with human-machine interfaces and job organization issues. As they point out 'most of these issues are not embedded in the nature of the technology per se' (Clegg and Symon, 1989: 37). In other words, in spite of the wishes and intentions of the social scientists involved in these projects, the technology was left to the technologists.

It is important that technologists should themselves develop an understanding of human issues, be able to investigate them, and determine what impact they have on system design. Attempting to achieve this by bringing them face-to-face with users in formative evaluation exercises has been one of the key objectives of the SEM. The result has been a significant transformation of the culture of the design team. It would appear that one of the main factors in the success of a 'human-centred' development project, such as PEN&PAD, is the perspectives and culture of the technologists involved. Social scientists can play an important role in challenging the traditional culture of system developers by transforming the technical development process into an explicitly sociotechnical process.

The key here is empowered user involvement, without which there would be little real pressure on system developers to behave differently.

It might be argued that while the iterative user-focused approach advocated here is fine for research projects, but that it is impractical and expensive in commercial and industrial development projects. The obvious reply here is that, in the long run, spending time and effort to understand the needs of users might be a lot less expensive than producing an unusable or unacceptable system. However, one important consideration is the social position of the end users. Empowered user involvement is perfectly feasible when dealing with doctors and other professional personnel. Yet, it is likely to pose severe difficulties in relation to clerical and manual workers, some of which are clearly illustrated in the manufacturing projects described by Badham (Chapter 6).

Knowledge representation and human-centred systems

Despite other forms of knowledge representation, rules are still mainly used in the development of expert systems. It has already been shown that because rule-based systems are designed primarily for automated reasoning they remove user involvement and control. Whilst rules are ideally suited to automated reasoning they have been shown to be of little use as a means of communicating knowledge.

In the case of the NUMAS, the AI/KBS idea of a semantic network was used to guide the development of parts of the hypertext system. This kind of application was not an expert systems but was it a knowledge-based systems? Gaines and Vickers (1988) argue that hypertext systems can be knowledge-based systems. In their view, the problem is that the AI/KBS community has redefined knowledge to mean that which can be codified for automated reasoning in a computer. On the other hand, the experience of this community has also been important in gaining an understanding of human knowledge and how it might be represented and structured. Some aspects of this understanding were used in the development of the NUMAS hypertext system, such as the distinction between procedural and conceptual knowledge, and a semantic network approach to knowledge representation. The resulting approach is aimed at providing the user with access to information and knowledge, making it an approach which is very much on the border between information systems and knowledge-based systems.

The Terminology Base developed in the PEN&PAD project is very definitely a knowledge-based system. The semantic network formalism has been designed specifically to provide the user with a medically sensible framework for data entry. Representing knowledge in the form of a semantic network has some similarities with the way it is understood that human memory is organized (Baddeley, 1990), However, it is not claimed that the GRAIL semantic network formalism is the basis of a generalized approach to cognitive modelling. To be useful, the GRAIL models of medical terminology within the Terminology Base must reflect aspects of the way in which doctors think about medical concepts. By way of an analogy the relationship between GRAIL models and the real world concepts they represent might be compared to the relationship between a stick person drawing and a photograph of a real person; the basic structure is represented in the stick person drawing but it does not possess the rich detail of a photograph. The accuracy of the GRAIL model will not be assessed by means of psychological experiments. Rather the test of the accuracy of the Terminology Base will be determined by how well it supports the data entry and other tasks of doctors using a computerized patient record and other medical information systems.

Conclusion

In one sense, AI/KBS technology poses exactly the same kinds of problems as the traditional technology-centred approach to mechanization and automation technologies. In the popular manifestation of AI/KBS technology in the expert systems paradigm, the purpose is to replace skilled and knowledgeable human action. This technology-centred purpose is encapsulated or enshrined in the technology. This is illustrated both in the technology-led medical expert-systems development cycle and in the inability of expert-systems technology to be adaptable to meet the requirements for knowledge-based advice giving proposed in the development of NUMAS. A redefinition of the purpose of the AI/KBS systems in terms of providing genuine decision support has led to the use of different aspects of AI/KBS technology. In different ways both the NUMAS and PEN&PAD projects have used the AI/KBS notion of a semantic network to represent, structure and store knowledge. Perhaps these projects point to an alternative role for the application and development of AI/KBS technology. Rather than attempting to emulate the way human beings think, a more useful application of AI/KBS technology might be the development of ways of representing and storing information or knowledge which correspond more accurately to the ways human beings represent and store information.

Where the purpose is to support the role and enhance the skill and knowledge of users, then empowered-user involvement in the design and development process is essential. The technology-centred view is deeply ingrained in the culture of technologists who consequently find it difficult to understand and accept the need for user involvement. Even where this need is accepted, technologists have little training or understanding of how to interact with users. In SEM, social scientists played an important role in facilitating a dialogue between users and developers. This not only contributed significantly to the development of the PEN&PAD system but also produced a major shift in the culture of the development team. The experience of SEM in the PEN&PAD project suggests that empowered-user involvement along with the participation of social scientists as facilitators is a useful mechanism for challenging the prevailing technology-centred culture of system designers and developers. However, empowered-user involvement may not be easy to achieve especially in traditional production environments where difficult political and industrial relations issues are raised. Nevertheless, there are some areas, such as the development of systems for doctors and other professional personnel, where real commercial pressures make approaches such as SEM serious and viable options.

References

Baddeley, A.D. (1990) *Human Memory: Theory and Practice*, Hove: Lawrence Erlbaum Associates.

Brown, P.J. (1989) 'Hypertext: dreams and reality', Keynote address at Hypertext II, University of York, August.

Churcher, P.R. (1990) 'Explanation in expert systems and pseudo-expert systems (Hypertext)', paper presented at British Computer Society HCI/Expert Systems Northern Group, Manchester, 14 March.

Clancey, W.J. (1983) The epistemology of a rule-based expert system — framework for explanation, *Artificial Intelligence*, **20** (3), 215–251.

Clegg, C.W. and Symon, G. (1989) A review of human-centred manufacturing technology and a framework for its design and evaluation, *International Reviews of Ergonomics*, **2**, 15–47.

Conklin, J. (1987) Hypertext: an introduction and survey, *Computer*, **20** (9), 17–41.

Coombs, M.J. and Alty, J. (1984) Expert systems: an alternative paradigm, in: Coombs, M.J. (Ed.), *Developments in Expert Systems*, London: Academic Press, pp. 135–157.

Corbett, J.M. (1985) Prospective work design of a human-centred CNC lathe, *Behaviour and Information Technology*, **4** (3), 201–214.

Gaines, B.R. and Vickers, J.N. (1988) Design considerations for hypermedia systems, *Microcomputers for Information Management*, **5** (1), 1–27.

Heathfield, H.A. and Wyatt, J. (1993) Philosophies for the design and development of clinical decision support systems, *Methods of Information in Medicine*, **32** (1), 1–8.

Kirby, J. (1991) 'Knowledge Based Systems in Computer Aided Control Systems Design', unpublished PhD Thesis, University of Sheffield.

Landes, D.S. (1969) *The Unbound Prometheus: Technological Change and Industrial Development in Western Europe from 1750 to the Present*, Cambridge: Cambridge University Press.

Laub, A.J. (1985) Numerical linear algebra aspects of control design computations, *IEEE Transactions on Automatic Control*, **30** (2), 97–108.

Rector, A.L., Horan, B., Fitter, M., Kay, S., Newton, P.D., Nowlan, W.A., Robinson, D. and Wilson, A. (1992) User centred development of a General Practice Medical Workstation: The PEN&PAD experience, in: *Proceedings of CHI '92*, New York: ACM, pp. 447–453.

Rector A.L., Kay, S. and Howkins, T.J. (1988) Facilitating the doctor-machine interface, in: Starling, P. (Ed.), *Proceedings of Current Perspectives in Health Computing*, Weybridge: BJHC Limited, pp. 213–222.

Rector, A.L., Nowlan, W.A. and Glowinski, A.J. (1994) Goals for Concept Representation in the GALEN project, in: *Proceedings of the Seventeenth Annual Symposium on Computer Applications in Medical Care (SCAMC)*, New York: McGraw-Hill, pp. 414–418.

Rosenbrock, H.H. (1977) The future of control, *Automatica*, **13** (4), 389–392.

Rosenbrock, H.H. (1985) 'Engineering Design and Social Sciences', paper presented to ESRC/SPRU Workshop on New Technology in Manufacturing Industry, May.

Rosenbrock, H.H. (1989) A flexible manufacturing system in which operators are not subordinate to machines, in: Rosenbrock, H.H. (Ed.), *Designing Human Centred Technology: A Cross-Disciplinary Project on Computer-Aided Manufacturing*, London: Springer-Verlag, pp. 177–184.

Shortliffe, H.E. (1976) *Computer-based Medical Consultations: MYCIN*, New York: Elsevier.

Taylor, J.H. (1988) Expert-aided environments for CAE of control systems, in: *Computer Aided Design in Control Systems 1988; 4th IFAC Symposium*, Oxford: Pergamon, pp. 7–16.

Taylor, J.H. and Frederick, D.K. (1984) An expert system architecture for computer aided control engineering, *Proceedings of the IEEE*, **72** (12), 1795–1805.

Taylor, J.H. and McKeehan, B. (1989) A computer-aided control engineering environment for multi-disciplinary expert-aided analysis and design (MEAD), in: *Proceedings of the National Aerospace Electronics Conference NAECON 1989*, New York: IEEE, pp. 1798–1806.

9

Will Symbiotic Approaches Become Mainstream?

Jos Benders, Job de Haan and David Bennett

Introduction

The individual contributors to this volume have each interpreted the umbrella term 'symbiotic approach' from their own perspective(s). Yet, the common denominator was found in the balanced combination of technical, social and organizational facets in order to design and implement properly functioning, and hence economically effective, production systems of all kinds. Compared with more traditional design approaches, this results in more attention for the interests and position of employees, not just as a goal in itself, but also, and at least equally important, in the firm belief that that is a way to improve a system's performance. And although this noble goal may sound self-evident and few people will object against it, reading the chapters makes one realize the profound difficulties incurred in its realization. The basic unitary intention of symbiotic approaches can easily become crushed in political, pluralistic processes (see Morgan, 1986: 185–194), which again underlines the importance of managing these hardly manageable processes. Furthermore, from a structural point of view the 'integral' character of symbiotic approaches makes them highly complex.

In the first chapter of this book, it was stated that:

> Although the importance of symbiotic approaches has been stressed repeatedly during the last 40 years as a prescriptive device, it is striking that descriptive studies time and again demonstrate that these prescriptions seem to have contributed little to the design of new production systems in practice. Thus, the diagnosis that technical systems fail to achieve their goals due to a lack of consideration for organizational and social issues can hardly be called new. Repeatedly, researchers have pointed to insufficient performance of production systems because social, organizational and/or human factors were ignored when the system was designed. Despite this repeated diagnosis, the same problems occur over and over again.

During the course of the project, statements in most of the contributions provided support for the initial impression that symbiotic approaches have, as of yet, only had a limited impact on production systems' design, which underlines the relevance of the questions posed in the introductory chapter. Why have symbiotic methods not realised their promising potential? What actions need to be undertaken to create more favourable conditions for the implementation of contemporary symbiotic approaches? In other words: how can symbiotic approaches become mainstream?

Furthermore, it was asked what were the basic differences and similarities between different symbiotic methods (in various countries), and was there anything 'new under the sun', or were these approaches basically the same? These questions are important to the issue of transferring approaches to other countries, as is explained later in this chapter.

The contributors' differing perspectives and interpretations mean that not all chapters address explicitly the questions posed. Yet, all have given their own answers, or pieces of answers, and it seems worthwhile to use these as building blocks for this final chapter. It should be stressed that the opinions expressed are not necessarily shared by every single contributor to this volume.

The answers are grouped into three categories:

1. the macro context;
2. structural characteristics;
3. political processes.

The latter two categories may be distinguished for analytical purposes but in the real-life they are two sides of the same coin, in other words: in any design process they are present simultaneously. Finally, the questions concerning the similarities in, and differences between, national symbiotic approaches are dealt with under the heading 'structural characteristics'.

Macro context

The macro context refers to all factors influencing the development and application of symbiotic approaches that are (normally) beyond the sphere of influence of the individual firm. It should be pointed out that the macro context is not necessarily confined to national boundaries, as Wobbe (Chapter 2) makes clear by discussing the supra-national role of the European Union.

The factors discussed here are, subsequently, the role of the government, the industrial relations system, the educational system and labour market, and the existence of infrastructural arrangements. However, many of these factors are interconnected; for instance, the government can intervene in the educational, industrial relations and infrastructural systems. Thus, the choice for a particular heading sometimes has an arbitrary element.

Despite the fact that the contributions stem from a variety of countries and although there is reason to believe that preferences for job design and employee participation are influenced by cultural differences (Hofstede, 1984), the theme 'culture' has hardly been addressed. Van Bijsterveld and Huijgen (Chapter 3) point to underlying cultural values in Dutch and Japanese conceptions of the quality of working life, but in the only chapter with an explicit cross-national focus, that by Wobbe, is the explanation for differences found in institutional rather than cultural factors. In order to avoid a culturalistic flavour in this text by falling in the trap of treating 'culture' as a 'black-box catch-all concept' (Lane, 1989: 29), i.e. as an *ex post* explanation for differences found, this issue is not dealt with as a separate item (see the section on Employees below).

Government

The European Union's research programmes have stimulated the development of various symbiotic technologies, for which an important role was foreseen in maintaining a competitive edge and in overcoming the failures of traditional production systems. As Wobbe points out, much missionary work needs to be done in communicating the knowledge of anthropocentric production systems (APS) because there is still a lack of awareness concerning their benefits. The emphasis in the 'action demand for public authorities' lies in diffusing knowledge of APS at various levels and to various parties, ranging from universities to individual firms, which is required to make the new Leitbild really 'tick'. Even in Germany, identified by Wobbe as the country with the most favourable prospects for APS, the traditional Leitbild has retained its dominant position. Generally, the need to diffuse the knowledge about symbiotic approaches is greater than the need to develop them further.

Latniak described that in Germany both the federal and some federal states' (Bundesländer) governments have stimulated the development of symbiotic approaches by developing and funding large-scale research programmes in which research institutions and firms cooperate closely. Probably more than in any other country, research projects were started up in a large variety of different industrial sectors, developing such specific technical systems as shop-floor-oriented programming systems (WOP) as well as design methodologies (STEPS). In the best cases, the funding concerns research programmes lasting several years after which the continuity may become endangered, while in the UK and the USA, the funding seems to be mostly on a project basis, making its continuity even more problematic and putting pressure on researchers to produce results quickly.

The German government also plays an important role in backing the power position of various parties but, most importantly, of that party which is generally weakest; the employees. Latniak (Chapter 5) refers to the German *Betriebsverfassungsgesetz* which ensures a strong position of workers' councils at various levels of co-determination on the level of the firm. In this way, they can influence decision-making, which remains, however, the legally defined domain of management.

The institutionalization of worker participation is by no means common, as Badham (Chapter 6) comments when discussing the lack of such backing in Australia. Furthermore, one may have doubts about the effectiveness of legislative backing of parties involved in symbiotic design processes because it does not seem likely that the much needed collaborative attitudes (see Wobbe and the section on Employees below) can be made obligatory.

Overall, governments at various levels can help create and sustain a climate and conditions that are favourable for the development and diffusion of symbiotic approaches, but they cannot enforce them.

Industrial relations system

Wobbe argues that the lack of collaborative industrial relations in countries such as France, Italy and the United Kingdom is a serious impediment to the diffusion of symbiotic approaches. Unions in countries with adversarial industrial relations may see cooperation with employers as 'collaborating with the enemy', an opinion which is often expressed by radical unions. In a similar vein, employers' associations may consider union involvement in job and organization design as an invasion into a domain which they consider their own.

The cooperation of Swedish unions and employers' associations stimulated the development and diffusion of symbiotic methods but, as Karlsson (Chapter 4) shows, other factors in such a cooperation may play a role, such as collective labour agreements which prohibited the use of wage increases to attract workers. This urged employers to find other ways of attracting employees, in this case by improving the quality of working life. Such an agreement helps to create an environment receptive to the use of symbiotic methods. It can be achieved because employees' and employers' representatives formulate collective labour agreements on their own and is probably more effective than government policies. However, such an agreement is the product of specific conditions at a particular point in time. In the Swedish case these conditions included an extremely tight labour market, which backs the power positions of the unions, and a government policy favouring harmonious industrial relations. As economic conditions worsened and the labour market scarcities were resolved, the situation may have become less favourable as the closure of Volvo's Uddevalla plant seems to indicate. However, as Karlsson demonstrates, other Swedish firms still pursue sociotechnical projects, which is perhaps an indication that such approaches become accepted after a long incubation period.

Overall, the prospects for acceptance of symbiotic approaches are better in countries with collaborative labour relations. In other countries, the change from adversarial to collaborative relations will be a necessary but difficult path. Here, management of individual firms will have to take the lead against the prevailing way of thinking in their macro environment.

Educational system and labour market

A highly skilled workforce is generally seen as an important prerequisite for the application of symbiotic approaches. For instance, Wobbe points to Germany's advantage in this respect compared with France and the UK. Because symbiotic designs stress the creation of jobs that are challenging and provide learning opportunities, so the demands on employees increase as well. In some countries, such as Spain, Portugal and the UK, there is a lack of sufficiently skilled personnel due to a relatively underdeveloped educational system. As the underutilization of employees' skills was one of the major reasons to start developing symbiotic approaches (see for instance the origin of the Dutch sociotechnical approach), this phenomenon seems rather remarkable; the policy to upgrade jobs in order to bring them into balance with employees' qualifications appears inconsistent with a situation where an insufficient number of employees possess such qualifications. The reason may be that symbiotic approaches originated in mass production environments, where from the point of underutilization of skills the need for such approaches is high; yet, many of these approaches were developed in manufacturing industries that traditionally provide higher-level jobs and where the underutilization was less of a problem. Latniak even mentions a possible bias in German approaches where the machine tool industry was important for their development. Symbiotic approaches have to be cautious about posing demands on employees that are too high. Whereas it is vital to bring the demand for and supply of jobs into balance, both in qualitative and quantitative respects, this does not mean *per se* that skill requirements posed by jobs need to be raised in all circumstances. The important issue is that a balance should be found; if not, this is likely to be an impediment.

Infrastructural arrangements

In many countries there are research centres which focus on the development, application and implementation of symbiotic approaches (see, for example, the contributors' affiliations), but no country seems to have vested institutions for promoting them. In other words, a comprehensive and generally accepted infrastructure to support symbiotic approaches does not exist, but is much needed, especially to support small- and medium-sized enterprises. It may also be a means to overcome the problems of acceptability, and hence cooperation, in the academic circles addressed by Latniak, and of the persistence of the traditional Leitbild in many professional communities (Wobbe).

Governments may take a role in stimulating such an infrastructure, but for it to have a lasting impact it should be able to stand on its own feet without governmental support. Institutions (such as educational centres and consulting firms) belonging to such an infrastructure should be able to finance themselves by acquiring projects from firms. Diffusion of knowledge about symbiotic approaches could then occur via these institutions' marketing efforts.

The lack of such an infrastructure contrasts sharply with the situation for more traditional approaches. An alternative to setting up a new infrastructure may be the reorientation of this existing infrastructure, a point brought in by Latniak. For instance, in Germany, the established institute REFA still dominates the design guidelines in industrial engineering, and few of the symbiotic thoughts have been incorporated into them. Majchrzak and Finley (Chapter 7) report that Scientific Management methods still have a strong influence in the USA. The Netherlands are a more positive case as insights from modern sociotechnology have been incorporated into the new labour law.

Structural characteristics

This heading is used to deal with those aspects that are not explicitly process-oriented, but instead concentrate on the result of such design processes. These are discussed in two sections, 'restructuring organizations', and 'the role of technical systems', after which the possibilities to exchange ideas internationally are examined.

Restructuring organizations

It has been stated that diffusing knowledge about symbiotic approaches is more important than developing them further. However, the latter is by no means trivial. Even in Germany, as Latniak stresses, there is no coherent, integral symbiotic design methodology, despite the need for one. At the same time he, and the other contributors, recognize that such an integrated approach would need to encompass a large number of aspects. The overall complexity is not only caused by the number of aspects included, but also by the interaction between these aspects. Majchrzak and Finley recognize this as one of the dilemmas of the sociotechnical paradigm. Perhaps one could even state that there is an inherent and irresolvable contradiction in a truly 'integral' approach; it is precisely its integrated nature that makes the approach so complex that it can hardly be managed. An integrated approach may be seen as desirable and is strived for as, for instance, the call to extend the Dutch, German and Swedish approaches show. Though in both practice and theory this is an unachievable ideal, its importance as a goal should not be underestimated. The ultimate, ideal, production system with which everybody involved is content may never be created, but its opposite can to a large extent be avoided, which is the design of production systems from a partial point of view. As it does not seem realistic to include all possible aspects and their interrelations *ex ante* in the design of a production system (as a truly integral approach would involve) the term 'integrative' seems preferable to 'integrated'. However, this should never be an excuse for reverting back to the old habit of creating partial designs.

In practice, the problem of including all relevant aspects in the production system's design may further be evidenced by the idiosyncrasies of many cases. A 'grand theory' would need to inform general guidelines which hold for all (or at least the majority of) cases, and at the same time take into account the specificities of the particular case in hand. If these specificities are overemphasized, which is not unlikely given the highly practical content of many examples, the theory becomes biased. A further danger already emerged in the American sociotechnical approach, namely its very abstract character which obscures its practical use.

There seem to be a number of possible courses of action to help to resolve these problems and these are prerequisites for the successful use of symbiotic approaches. One of these is even called 'ACTION', which seems to be an extremely powerful tool to reach a well-considered design. This software package makes it possible to include a large number of variables, and to assess the impact of specific variables on others. Latniak proposes a different solution, namely a modular approach which can be adapted to organization-specific circumstances but incorporates some basic principles. Yet another possibility lies in the Dutch sociotechnical approach, which follows a 'logical design sequence', i.e. a fixed number of steps in a fixed sequence (although, in practice, design processes are not so linear than this theory prescribes).

Starting with the characteristics of the products that have to be produced (in terms such as product variety and batch sizes), and thereby the necessary production tasks, the latter are grouped by the design steps of paralleling and segmenting, into more or less natural entities, those parts of the overall production process that from a production point of view form a logical whole. The responsibility for managing these wholes are then assigned to so-called 'whole-task groups', to which as many regulatory tasks are assigned as possible. In this way, modern sociotechnology provides a means to create an environment in which semi-autonomous teams can flourish. Although modern sociotechnology needs further development, as van Bijsterveld and Huijgen state, its strong point is that it focuses on the organization structure (layout), because it is this structure that provides a semi-permanent setting in which production takes place. Whereas the layout is also mentioned in most of the other contributions, its importance is recognized most explicitly by modern sociotechnology, which places it at the very heart of design methodology. However, it must be recognized that modern sociotechnology is by no means unique in this respect, as immediately becomes clear when reading Badham's description of team based cellular manufacturing as used in Germany and Australia.

As well as addressing concerns for the quality of working life, the necessity of applying a symbiotic approach such as modern sociotechnology was motivated by changing product markets with higher quality demands, shorter product life cycles and batches, continued price pressure, and, most notably, a larger product variety. Following a design logic such as modern sociotechnology, these changing market demands provide a stimulus for the creation of meaningful jobs. The 'new' organization structures designed to satisfy these demands are not likely to function properly with the traditional unskilled jobs. However, the conclusion that the quality of working life will necessarily improve as a result of changing market demands is too optimistic. Wobbe points to the continuing importance of standardized mass production in many countries, and when this becomes less standardized the situation can be characterized as 'repetitive manufacturing' which is ideal for lean production with its line structure (van Bijsterveld and Huijgen). Furthermore, Latniak points to the fact that market demands and organizational structures are only 'loosely coupled', indicating degrees of freedom that may work out positively or negatively for the quality of working life.

The role of technical systems

It is perhaps rather surprising that the role of technical systems came out less prominently than anticipated ('technical' here refers to physical entities such as machines, not to the grouping of machines as is generally the case in the somewhat confusing term 'sociotechnical'; see van Bijsterveld and Huijgen). Wobbe and Latniak point out that the shop-floor programming package WOP does not guarantee any favourable impact on the quality of working life, which fits well with Badham's notion of 'local configuration'. WOP must be seen as a factor which makes possible the creation of skilled 'integrated' jobs. It differs from many traditional programming packages which are quite complex to programme, and thus pose high intellectual requirements which in turn form a 'natural barrier' to the creation of integrated jobs. This effect of the technical complexity of conventional programming systems on job design may not be intentional, but it occurs nonetheless, which is an excellent example how the unforeseen interaction of variables may have negative consequences. It also shows that there is no technological (or preferably 'technical') determinism. Generally, a reference to technical determinism is made to point out that low quality jobs are by no means a necessity caused by a specific technical system. However, it is logical the reverse also holds. For the same reason that degrees of freedom exist a symbiotic technical system does not guarantee high quality jobs. Badham's emphasis on the configurational process, in which a specific technical system is embedded in the design of a production system, can be seen in this light. The point also holds for Kirby (Chapter 8) who recognizes that many of the issues which are the domain of social scientists are not embedded in the technology, so that 'technology was left to the technologists'. Although this is not necessarily a problem in view of the above discussion, there are cases in which organizational principles materialize in the design of technical systems, such as the case of production planning and control systems, or even information systems in general (see Kirby). Such cases therefore restrict freedom of design. Kirby's remark that 'It is important that technologists should themselves develop an understanding of human issues, be able to investigate them and determine what impact they have on system design' can therefore only be underlined, especially for relatively new technical systems, such as knowledge-based systems, where the danger of falling into the old technocentric trap seems greatest. In the early development phases technicians have a monopoly, which may easily result in negative organizational or social consequences in later phases. This may happen unintentionally as stated above, but as Badham points out when referring to the deeply ingrained engineering view of humans as unreliable, in other cases it may also be an explicit goal.

International exchange of ideas

Partly because of the cross-national differences outlined in the section 'Macro content', but perhaps more importantly because of the language barriers, the exchange of information between countries is more limited than seems desirable. As Latniak points out, an impressive number of often substantial projects in Germany have progressed throughout the last 20 years in relative isolation. Knowledge about the development and application of symbiotic approaches might accumulate faster if researchers' access to information about the progress in other countries was less of a barrier. One reason for the prominent role that small countries, such as Sweden and The Netherlands, play in this area may be that, being small, they are open to foreign influences. For instance, British, German, and especially Swedish influences can clearly be tracked down in the history of Dutch modern sociotechnology, although they are not always recognized explicitly. In the Dutch, German and Swedish contributions the impact of 'lean production' on the development of their countries' approaches is mentioned, indicating a lively interest in learning from other countries. The interest managers displayed for lean production is almost infectious, and contrasts sharply with the attention for symbiotic approaches, which can be explained by lean production's stress on economic performance whereas symbiotic approaches are noted in the first place for their social performance (van Bijsterveld and Huijgen).

In spite of the seeming isolation in which symbiotic approaches have been developed, the similarities between the approaches adopted in different countries is often striking. For one thing, team working scores high on the list of preferred solutions, although the terms given to approaches that enhance the creation of teams differ. Germany's '*Produktionsinsel*', Dutch 'modern sociotechnology', Australian 'team-based manufacturing cells', the Swedish 'teams', and the common recommendation of using teams in American sociotechnical projects, all point to the worldwide popularity of the team concept. However, beneath the surface, considerable variation in team concepts used may be found. For instance, Dutch whole-task groups may show considerable differences when compared to American teams. Another striking similarity can be found in the German VERA/RHIA and the approach of Dutch modern sociotechnology with respect to work content.

Similarities may stimulate the international transfer of approaches, but at the same time it may never be clear whether a specific phenomenon has a foreign or an indigenous origin. A case in point relates to lean production, which originates in Japan. One of its characteristic aspects is team working and, although teams in Japan bear only a remote resemblance to the semi-autonomous teams proposed by sociotechnical approaches, the latter may be introduced under the banner of lean production (see Latniak; van Bijsterveld and Huijgen). In a similar vein, if German-Australian team-based cellular manufacturing has considerable overlap with Dutch modern sociotechnology, how can each of their influences be identified in, for instance, a Latvian company which has taken notice of both approaches before implementing cells and work teams? Whereas such issues may seem primarily to be of academic interest, and only of secondary importance to many practitioners, they are vital for establishing the possibilities for an international exchange of ideas.

Political processes

If one theme dominated the discussion during the Workshop, it was politics, especially politics within the organization as the German word *Mikropolitik* makes clear. Symbiotic redesign projects 'are infused at their very core with power relations, organizational conflict and cultural discord' (Badham), affecting the interests of all parties involved and 'contesting the terrain'. Partly depending on their power position, parties are successful in influencing the outcome of the process in their own interest. As the descriptions of the projects by Badham, Kirby, and Majchrzak and Finley make clear, it is hard to distinguish between the design and implementation phases in real life.

A discussion about micro-politics may start by distinguishing three parties that each have their own power positions and interests:

1. management;
2. employees;
3. designers, subdivided into social scientists and engineers.

Management

Management is, in the first place, interested in the economic performance of a production system, and management will ultimately take the final decision about a system's adoption (see the examples in Badham's contribution). Even in Germany where employees' codetermination rights are legally anchored, decision-making remains a managerial prerogative. Therefore, one of the main challenges for promoters of symbiotic approaches is to provide hard proof that these approaches are, economically speaking, superior to conventional design approaches. If this can be done, and provided that this proof is accepted by managers, the highest barrier to symbiotic approaches becoming mainstream has been climbed.

Whereas symbiotic approaches claim considerable economic benefits (see Arthur, 1994, for an analogous case concerning human resources management) managers do not seem convinced, as can be read in almost every single contribution to this book. Where symbiotic approaches have in practice paid single-sided attention to social benefits, and the names of many of these approaches express a primary interest in social aspects, managerial scepticism will be especially hard to overcome. The importance of solid investment calculations has been acknowledged and much research has gone into economic evaluations of design projects. However, star cases may be over represented (see Latniak and Karlsson). Moreover, partly because of the complexity of many redesign projects and the uncertainties inherent in them, the conventional problems of capital budgeting remain present. However, demonstrating a system's economic performance is not always sufficient to overcome managerial scepticism, as Badham demonstrates when a more conservative option was chosen in spite of the cost–benefit analysis favouring a cell layout. As Majchrzak and Finley quote of a manager 'If you don't feel STS in your heart, you can't be sold on it'.

Employees

Employees' interests can be subsumed under the title 'quality of working life', which is an umbrella concept. Van Bijsterveld and Huijgen refer to the social performance of production systems, splitting it up into:

- work content;
- work environment;
- terms of employment;
- labour relations.

The emphasis within most approaches lies in work content, sometimes even to the neglect of the other aspects. For instance improving work content, and thus raising skill requirements, is a central concern in Sweden. Karlsson discusses the increased autonomy of work groups, and mentions a substantial number of areas for which shop-floor workers can take over responsibility. While academics generally see such a development as advantageous for workers, the latter do not necessarily share this view, an objection raised by van Bijsterveld and Huijgen.

Measures concerning the work environment involve improving the physical conditions and ergonomics in general. For adherents of symbiotic approaches these seem to be basic preconditions which are so self-evident they are hardly dealt with explicitly. However, even in an advanced industrial nation there may be considerable room for improvement in this area, as the case of the Japanese car manufacturers mentioned in the introductory chapter proves.

Terms of employment include, among other things, wages (and other forms of financial compensation), which do not always receive the attention they deserve. Latniak points to the fact that these often fall outside the scope of responsibility of project groups. He also points to the complexity of designing reward systems that both enhance group work and stimulate individual commitment. In a similar vein, Karlsson mentions the problem of increasing stress caused by certain productivity-related wage systems on the one hand and of weak incentive structure as an impediment on the other. Perhaps a simulation program such as ACTION may help to overcome this neglect of wages as part of a redesign process.

Regarding labour relations, participation features prominently in many symbiotic approaches and may fulfil several functions. In the first place, there is the concern for democratization which comes to the foreground most dominantly in Gustavsen's *Democratic Dialogue* approach (Karlsson; van Bijsterveld and Huijgen). Although democratization as a goal in itself was an important stimulant for the early development of some symbiotic approaches, the emphasis seems to have shifted to other functions of participation. Second, input from employees is needed in order to design production systems and, in later phases, to enable them to run smoothly and effectively. As Kirby, in particular, makes clear, user input in the design of information systems is indispensable, a fact which is increasingly recognized (see Leonard-Barton and Sinha, 1993). In a manufacturing environment, user input is equally important, as illustrated in the local configuration process. Once put in place employee participation remains an essential feature expressed in the form of acting autonomously to run work processes. It is stressed in modern sociotechnology that employees should be given 'regulating capacity' because they are in the best position to intervene in case of disturbances. A third function of employee participation is gaining employee acceptance of a production system by taking into account their wishes and preferences. The fourth and final function can be complying with legal requirements. This form hardly deserves the name 'participation', but may occur where the merits of participation are not acknowledged by management.

The power position of employees is weak relative to that of management in that the latter decides to what extent the former may participate. In certain circumstances, however, the position of employees, or future users, can be fairly strong. In the first place their knowledge about production processes is a power base. First, if employees are not willing to share this knowledge with designers, the performance of production systems will probably be sub-optimal. Second, as Kirby points out, the social position of future users influences the feasibility of user involvement. Professionals such as doctors are more likely to put their stamp on a project than lesser skilled manufacturing employees.

Finally, employees may be supported by representative bodies such as workers' councils or unions, whose right to participate may be legally backed and enforced.

Many design-oriented approaches, such as the German VERA/RHIA, are directed only at improving the quality of working life. By definition symbiotic approaches stress the quality of working life, but at the same time they place it in perspective with other goals to be achieved. Nevertheless the impression has arisen that in many instances symbiotic approaches were used to improve the quality of working life, neglecting other goals. For instance, when Karlsson stresses the emphasis on job enlargement, team work and a democratic leadership style in Sweden. Also, when van Bijsterveld and Huijgen mention modern sociotechnology's acknowledged improvement of 'social performance', this may easily lead to the false impression that the goal of symbiotic approaches is constrained to the amelioration of job dissatisfaction (compare with the re-labelling of the German federal research programme *Humanisierung des Arbeitslebens* to *Arbeit und Technik*).

Designers

Designers are broadly divided into social scientists and engineers. They have to work together, and there is the need to cooperate with managers and employees in design projects. The mutual cooperation of academics is not without problems. As Latniak points out, mainstream academics still find crossing the borders of their 'home' disciplines hard to accept, which provides a problem for symbiotic designers who seek and need recognition in their own professional community. This mirrors the functional structure of many organizations, which is institutionalized in educational systems, where few students are exposed to other disciplines. This in turn is an important reason for the communication difficulties between social scientists and engineers, and the technologists' lack of understanding of human issues which was discussed by Kirby. In the same spirit, however, social scientists must overcome their fear of technical topics.

The main point of tension in the cooperation of designers on the one hand, and managers and employees on the other, concerns finding a balance between bringing in expert knowledge and letting other parties participate, a topic which comes strongly to the fore in van Bijsterveld and Huijgen's discussion of modern sociotechnology which they criticize for being essentially an expert approach. This is because it prescribes a fixed number of steps to follow, which do not always leave room for employee participation. An elegant example of cooperation is provided by Kirby's description of the PEN&PAD project, where social scientists facilitated the dialogue between users and the design team in a way that the designers were able to take technical decisions based on a solid understanding of the users' tasks and information requirements. The German STEPS method also stresses the interactive process between designers and future users into account, so that user knowledge can be incorporated into the software program under development.

Process of change

Ultimately all parties interact with each other in the change process which ranges from designing and building, to implementing, running and possibly adapting a production system. The problem of methods to structure change processes is that they seem to be attempts to manage the unmanageable. However, as Latniak points out, they cannot capture the change process in all its details, nor can the process be divided into manageable steps. Instead, there can be 'reference lines' as in STEPS or milestones, without that these are given an absolute value. Unforeseen factors and developments may emerge at any time and exert a considerable influence. Even stronger than is the case for design methods for the structure of production systems, the main task of process design methods being to provide guidelines rather than a blueprint for managing the change process. An illustration can be found in the 'highly iterative' supportive evaluation methodology (Kirby).

One of the most important aspects here was mentioned by Karlsson, which is 'Who defined the problem?'. More often than not the answer is 'management', leading Majchrzak and Finley to comment that the mission and the focus of the design team are often defined by management in a top-down process. Although there is consent about the importance of goal agreement before starting a design project, management's dominant position in setting goals may be an impediment to the formulation of a common perspective about what the project should achieve. The political nature of the change process clearly comes to light in an activity such as goal setting and in employee participation. As Latniak states, 'participation may lead to the modification of managerial goals'.

In a similar fashion, the Gothenburg model, as described by Karlsson, stresses the 'interest group analysis' during which differences are made explicit. In general, the contribution of process-design methods seems to be of bringing differences out into the open so they can be discussed and thereby problems resolved. Social scientists can perform the role of facilitators, making fruitful communication between parties possible, although such communication cannot guarantee a successful project. Even though communication difficulties between, for instance, employees and technicians may be resolved, it takes more than just communication to deal with conflicting interests

Given the extensive investment in money and time demanded by symbiotic design processes and the often troublesome course design projects take, it is no wonder that doubts arise concerning their necessity. As several contributors have pointed out, these are substantial barriers to their application, and reactive strategies are likely to continue to exist. The difficulties incurred in symbiotic design projects form an additional reason for the need to demonstrate that such projects lead to superior designs.

Conclusions

Symbiotic approaches still have a long way to go, but at the same time it cannot be said that there has been no progress. Many traditional assumptions concerning the design of production systems are no longer automatically taken for granted and, as all the contributions in the various countries show, symbiotic approaches have been developed and applied in real-life situations. However, their wide-spread diffusion cannot simply take place within a few decades. Traditional design approaches and *Leitbilder* have taken hundred of years to develop, during which they were firmly established and institutionalized. It would be an illusion to think that this development, with its far-reaching consequences, can be reversed in just a few years. The complexity and deeply political nature of symbiotic approaches are an additional barrier to their diffusion. Symbiotic approaches still are in their incubation period. Perhaps a hopeful sign is that under the banner of buzz words such as 'empowerment' and 'de-layering' many organizations have undergone restructuring processes, the outcomes from which share many familiarities with those sought by the more fundamental symbiotic approaches. The effect of projects inspired by such popular buzz words may not always be applauded. The mere fact that they have become popular indicates that the traditional neglect of managers for the human factor has encountered competition.

There is still a long road to travel if symbiotic approaches are to become mainstream. Priorities in their future development should concentrate on at least three areas:

1. the process of change with its inherently political character;
2. the diffusion of knowledge about symbiotic approaches;
3. the extension of knowledge about the content of symbiotic approaches.

Several methods are used to guide the change process and it cannot be structured in detail. What is needed on the part of all parties involved in a change process is the explicit recognition that such a process is political and that defending one's own interests is a valid way of acting. For one thing, this requires an appreciation of the interests of other parties, which should be brought out in the open and not be left implicit or hidden.

The diffusion of knowledge should be supported by infrastructural arrangements, although the best these can do initially is attempt to make the environment more receptive to the adoption of symbiotic approaches, because the ultimate decisions are taken in individual firms. If researchers can succeed in convincingly demonstrating the superior economic performance of symbiotic approaches, the creation of such an infrastructure may occur spontaneously. Making the results of such research and the content of symbiotic approaches easily comprehensible for the public at large will be an important step in the essential task of changing the traditional assumptions held. Governments may stimulate the creation and sustenance of such an infrastructure, as is already the case in some countries, but, in the long run, the institutions making up the infrastructure must be able to function without governmental support.

Further development of symbiotic approaches is also necessary. The basic principles seem to be known but several contributors have pointed out neglected, yet vital, areas in their countries. A software program such as ACTION may help to manage the complexity involved in extending what are already complex approaches, and in giving the parties involved better insight into the consequences of alternative designs, thus facilitating the change process.

The main questions posed in this book cannot be answered with a simple 'yes' or 'no', as should be clear now that its end has been reached. Hopefully the ideas and opinions expressed will nourish the flame of symbiotic approaches.

Acknowledgements

The authors are indebted to Mark van Bijsterveld and Erich Latniak for their comments on an earlier draft of this chapter.

References

Arthur, J.B. (1994) Effects of human resource systems on manufacturing performance and turnover, *Academy of Management Journal*, **37** (3), 670–687.

Hofstede, G. (1984) The cultural relativity of the quality of life concept, *Academy of Management Review*, **9** (3), 389–398.

Lane, C. (1989) *Management and Labour in Europe: The Industrial Enterprise in Germany, Britain and France*, Aldershot: Edward Elgar.
Leonard-Barton, D. and Sinha, D.P. (1993) Developer-user interaction and user satisfaction in internal technology transfer, *Academy of Management Journal*, **36** (5), 1125–1139.
Morgan, G. (1986) *Images of Organization*, Beverly Hills/Newbury Park/London/New Delhi: Sage.

Notes on Contributors

Richard Badham (PhD in Sociology, University of Warwick, 1983) is an associate professor and director of the Management of Integrated Technical and Organizational Change (MITOC) action research unit at the University of Wollongong, Australia. He is on the International Editorial Board of the *International Journal of Human Factors in Manufacturing, AI and Society*, and *Journal of Industry Studies*.

Jos Benders (PhD, University of Nijmegen, 1993) is an assistant professor at Department of Business Studies of the University of Nijmegen, The Netherlands. His research interests include the intra-organizational division of labour and the impact on this of technical systems.

David Bennett (PhD, Birmingham University, 1978) is head of the Technology and Innovation research programme at the Aston Business School, Aston University, United Kingdom. His research interests include issues relating to production systems design and technology transfer.

Mark van Bijsterveld (Masters Degree in Psychology, University of Leiden, 1992) is a doctoral candidate at the Department of Business Studies of the University of Nijmegen, The Netherlands. His research interests include information systems design processes, organizational politics and new production concepts.

Linda Finley (BA, Texas A&M University, 1984) is a licensed industrial engineer and an industrial engineering supervisor for Texas Instruments (USA). Her research interests include organizational design and plant layout.

Job de Haan (PhD, Tilburg University, 1983) is an associate professor at the Department of Business Administration of Tilburg University, The Netherlands. His research interests include organization structure and control, with an emphasis on information technology and cost-benefit analyses.

Fred Huijgen (PhD, University of Nijmegen, 1984) is a full professor of Business Administration at the Department of Business Studies of the University of Nijmegen, The Netherlands. His research interests include the intra- and inter-organizational division of labour, human resources management and labour market developments.

Ulf Karlsson (PhD, Chalmers University of Technology, 1979) is a full professor at the Department of Operations Management and Work Organization at Chalmers University of Technology, Gothenburg, Sweden.

John Kirby (PhD, University of Sheffield, 1991) is a Research Fellow in the Medical Informatics Group, Department of Computer Science, University of Manchester, United Kingdom. His research interests include design and development methodologies and technology transfer.

Erich Latniak (Dr rer soc FernUniversität/Gesamthochschule Hagen, 1994) is a member of the scientific staff of the Production Systems Department at the Institut Arbeit und Technik (Science Center North Rhine-Westphalia), Gelsenkirchen, Germany. His research and consultancy focus is on work organization, human resources management, and vocational training.

Ann Majchrzak (PhD, University of California at Los Angeles, 1980) is an associate professor of human factors at the University of Southern California's Institute of Safety and Systems Management and Industrial and Systems Engineering Department, United States of America. Her research interests include sociotechnical systems design; integration of technology, organizational and people systems design; macro-human factors; advanced manufacturing technology.

Werner Wobbe (PhD, Göttingen University, 1981) was head of the subprogramme 'Technology, Work and Employment' and 'Future and Industry in Europe' in the Research, Technology and Development Directorate of the Commission of the European Communities, Brussels: Belgium. His research interests include the social and economic shaping of technology development and its application, as well as its relation to various European industrial cultures.

Author Index

Subject Index